Police Powers 1

Police Powers I

PAUL F. MCKENNA

President, Public Safety Innovation, Inc.

With a foreword by The Honourable Fred Kaufman, C.M., Q.C.
Former justice of the Quebec Court of Appeal

Toronto

National Library of Canadian Cataloguing in Publication Data

McKenna, Paul F. (Paul Francis), 1952–
    Police Powers I

Includes index.
ISBN 0-13-040696-1

1. Police power—Canada. 2. Criminal procedure—Canada.
3. Canada. Canadian Charter of Rights and Freedoms. I. Title.

HV8157.M4 2001   345.71'052   C2001-900472-9

ISBN 0-13-040696-1

Vice President, Editorial Director: Michael Young
Editor-in-Chief: David Stover
Senior Editor: Sophia Fortier
Developmental Editor: John Polanszky
Marketing Manager: Sharon Loeb
Production Editor: Susan Adlam
Copy Editor: David Handelsman
Production Coordinator: Trish Ciardullo
Art Direction: Julia Hall
Cover Design: Dave Mckay
Cover Image: Photodisc
Page Layout: Carol Magee

1 2 3 4 5   06 05 04 03 02

Printed and bound in Canada.

Dedicated to Denny Hurley
Failed by Justice; Sustained by Friends & Faith

CONTENTS

CHAPTER FOUR
INTERIM RELEASE OF ARRESTED PERSONS 92

CHAPTER FIVE
POLICE DISCRETION IN CANADA   118

APPENDIX

FOREWORD

As Paul McKenna points out in the Preface, this book is designed to help the reader develop "a comprehensive understanding of existing police powers in Canada." And indeed it does. It is a welcome addition to the growing literature available in Canada for students preparing for a career in police work—an area much neglected in the past. But it goes further: it is a useful text for everyone engaged in law enforcement, clearly written, comprehensible, concise, yet accurate and complete.

Almost 20 years have passed since the *Canadian Charter of Rights and Freedoms* became a key constitutional document, and many of the dire predictions of mass acquittals of criminals made at the time—reminiscent of what was said when the *Miranda* rules were first laid down in the United States—proved to be needless fears. True, the Charter (as did *Miranda*) helped free a number of accused, but it also helped to bring about a higher standard of police work. And that was all to the good. As the author points out, the Charter "allows the courts to determine if the police are administering justice in a manner that is acceptable within the existing Canadian constitutional framework." If they are, their efforts will pay dividends in court; if they are not, they will have failed in their duty to uphold the law and society will be the poorer for it.

It is one thing to enunciate these ideals, and quite another to teach them to present and future police officers. The rules are not always clear, the law is sometimes confused, legal decisions may contradict each other, yet we expect a peace officer to make split-second decisions which will stand not only the test of time, but also subsequent judicial scrutiny. This is a problem not always appreciated by the public, and I am pleased to note that the author meets this head on in Chapter 5. As he notes, "discretion tends to place a significant reliance on the critical judgment of front-line officers as they function in an environment that is exceedingly self-directed." This is not easy, and while much will be learned on the job, good teaching at the outset will lay the foundation for better decisions in the field. This book will help to lay this foundation.

To the serious student this book is a gold mine of information. It is up-to-date—DNA samples, body impressions, telewarrants—all in the text, together with questions for consideration and discussion, case law, and references. Each chapter sets out the learning objectives, avoiding a trip into uncharted waters. It is, in short, an excellent tool for both teacher and student.

Recent events in Canada—*Marshal, Milgaard, Morin,* and others—underscored the need for better training of police officers. But even the best-trained and best-intentioned officers can only function effectively and efficiently if they have the support of their superiors, and that, regrettably, isn't always the case. Then, too, we have what has been called the "police culture." This is hard to define, but in essence it means that "we must stand up for each other, right or wrong." It also means "winning at any cost," and sometimes this leads to inexact and misleading statements in court, called "noble cause corruption." This is unethical and wrong, and clearly points to the need for better fundamental training.

As Justice Louis Brandeis once said: "The greatest dangers to liberty lurk in insidious encroachments by men of zeal, well-meaning, but without understanding." It is the purpose of this book to instill this understanding in students who may one day face these dilemmas.

The Hon. Fred Kaufman, C.M., Q.C.
Former justice of the Quebec Court of Appeal

Toronto
March 5, 2001

PREFACE

This textbook is intended to provide students with a resource for developing a comprehensive understanding of existing police powers in Canada. It is vital that a solid knowledge is acquired in this area, but also that individuals realize that the state of the law is subject to change. Therefore, the presentation provided in this volume will be useful in preparing the student to learn about the development of Canadian law with respect to police powers, but more significantly, to know where to turn in order to update and refine that knowledge. Canadian law is subject to change from two main sources: the legislature (federal and provincial) and courts. It is through updating these two fundamental sources that students will be able to ensure that they are aware of the most current law as it pertains to search and seizure, arrest, interim release, and other related topics.

The first chapter deals with the *Canadian Charter of Rights and Freedoms* and its impact on police powers in Canada. It traces the origins of the Charter as a keystone in the country's constitutional law. Also, this chapter looks carefully at the seven sections of the Charter that most impact on policing in Canada and considers relevant cases that have provided commentary and content on those sections.

Chapter 2 examines police powers of arrest and release in Canada. Working definitions of "arrest" and "detention" in Canadian law are set out. Arrest powers with or without a warrant are considered in the context of relevant Canadian case law. The concept of "telewarrants" is considered and discussed. A listing of general powers of arrest for a peace officer in Canada is provided, as well as the powers of arrest of anyone in the country.

Chapter 3 deals with the powers of search and seizure in Canada. Again, the Charter will be considered as it deals with this topical area. The concept of unreasonable search and seizure will be addressed as it has been established in Canadian law. There is also a discussion of the search warrant document and the procedures to be followed in acquiring such a document. This chapter further deals with the general powers of search under the *Criminal Code of Canada*. Warrantless searches are dealt with in some detail. The practice of using video surveillance, tracking devices, and telephone recorders is considered in the context of current Canadian law. Also, the new area of securing bodily substances for DNA analysis is reviewed. This chapter also looks at the case law as it pertains to the search of the media, search of the person, power of detention during a search, and the seizure of goods found during a search. Furthermore, this chapter will introduce the student to the powers of search under federal statutes other than the *Criminal Code of Canada*, as well as any powers of search under provincial statutes.

Chapter 4 deals with the interim release of arrested persons and includes a consideration of the application of statutory authority of such persons by patrol officers, supervisory officers, and a justice.

Chapter 5 examines the topic of police discretion in Canada and considers the implications for the delivery of policing services in the country.

This textbook includes relevant cases in all topical areas as well as questions for consideration and discussion in each chapter. There has been an effort to provide students and instructors with relevant related activities that can be adopted in order to consolidate learning in each chapter and to extend the understanding pertinent to individual topics discussed throughout this publication. Finally, an appendix containing the complete text of the *Canadian Charter of Rights and Freedoms* has been included for convenience.

It is important to emphasize the need for the student to confront some of the difficult legal language to be found not only in the statutes, but also in the judicial writings and decisions. While there has been considerable effort made to explain the substance of the relevant statutes and court decisions in ordinary language, it must be recognized that laws are drafted according to certain standards. Also, judges and lawyers acquit themselves of their legal responsibilities in a professional language that is not always easy for the layperson to understand. However, the nature of policing is such that those wishing to function in this field must become relatively conversant with the mode of expression used by their colleagues in the criminal justice system.

It is hoped that this textbook will provide a strong foundation for students to learn about basic police powers in Canada and to gain the skills necessary to research case law and statute law in a manner that will keep them current and well-informed.

ACKNOWLEDGMENTS

This book has grown out of a professional interest in policing in Canada that is now more than 20 years old. Over that time, I have had the privilege of learning from many individuals who were perceptive practitioners of the "art" of policing, as well as, several academic observers of the "craft" of policing. Both categories of colleagues have guided me toward a deeper understanding of what policing means, and what it might become. It has been valuable to work inside the network of Canadian policing, not as an officer, but as someone who functions to support and enhance the difficult work undertaken by these organizations. It has become strikingly clear that much needs to be done to ensure that the right kind of educational tools are made available to all levels of policing if innovation, continuous learning, and professional excellence are to be achieved and sustained.

Given the length of my tenure as a "student" of policing in Canada, it is impossible to recite the names of all my teachers in this discipline; however, the individuals listed below represent the most important sources of insight, inspiration, and integrity. They are listed in alphabetical order, along with their professional affiliation:

- William J. Closs, Chief of Police, Kingston Police
- Dr. Frum Himelfarb, Royal Canadian Mounted Police (Ottawa)
- Dr. Barb Juschka, Calgary Police Service
- Robert Lunney, Chief of Police (retired), Peel Regional Police Service
- Thomas B. O'Grady, Commissioner (retired), Ontario Provincial Police
- Clifford Shearing, Centre of Criminology, University of Toronto
- Christine Silverberg, M.A., APR (Chief of Police, retired), The Silverberg Kennedy Group, Inc.
- Philip Stenning, Centre of Criminology, University of Toronto
- Paul Wilhelm, Superintendent (retired), Ontario Provincial Police

I am also deeply indebted to Anne Dwyer, Centre for Continuing Education, McMaster University, for providing me with the opportunity to teach people who are striving to learn more about being and/or becoming police officers. It is a great gift to be in the presence of those who thirst for understanding and it has allowed me, as a writer, to test ideas in that important laboratory: the classroom. I also thank Mike Weaver, Media Relations, Kingston Police, for his generous efforts in providing photographs for this publication. It is a great pleasure to recognize the strategic vision of David Stover, Pearson Education Canada, who encouraged me in tackling this project, as well as his colleagues: John Polanszky and Susan Adlam for their professional management of this file. Also, David Handelsman is to be thanked for proving the infinite value of a competent copyeditor. Finally, I acknowledge the spiritual support of my partner, Lee Simpson, and the cerebral example set by my daughter, Kathleen.

CHAPTER ONE

For to root out Fraud,
prevent Violence and
Oppression, and to
preserve Peace and good
Order, are the most grateful
Pursuits of a good Heart
and an ingenious Mind.

John Fielding (1758)

THE CHARTER AND ITS IMPACT ON POLICE POWERS IN CANADA

LEARNING OBJECTIVES

1. Identify the origins of the *Canadian Charter of Rights and Freedoms*.
2. Outline the significance of the Charter for Canadian constitutional law.
3. Provide details on the relationship of the Charter to policing in Canada.
4. List the seven sections of the Charter most relevant to policing in Canada.

INTRODUCTION

This chapter will look at the *Canadian Charter of Rights and Freedoms* (the Charter) and consider its importance to policing in Canada. As a key constitutional document, the Charter represents a major accomplishment in the evolution of Canada, but it has also had a profound impact on the way in which policing is provided in this country. Exactly how the Charter influences the day-to-day workings of the police is the subject of this chapter. Why do police need to understand the Charter and its implications? The simple answer is that this legislation provides the tools for judges and the courts to assess the overall policies and procedures of police departments, as well as the individual actions of police officers as they function in their daily routines. The Charter is nothing short of the standard by which the courts measure whether the police are right or wrong in their approach to law enforcement. The Charter allows the courts to determine if the police are administering justice in a manner that is acceptable within the existing Canadian constitutional framework.

In this chapter, we will first look at the origins of the Charter in Canada and briefly trace some of its history. Next, there will be a consideration of the importance of the Charter for police officers in Canada. Finally, this chapter will provide a detailed listing of the seven Charter sections that most impact on police organizations and their members.

ORIGIN OF THE CHARTER

April 17, 1982, is an important date for policing in Canada. This is the date when the *Canada Act* came into force. Schedule B of that legislation is the *Constitution Act, 1982*, and Part I of the *Constitution Act, 1982*, is known as the *Canadian Charter of Rights and Freedoms* (the Charter). The Charter is of considerable importance to law and policing in Canada and should, therefore, be clearly understood. In combination with the *Canadian Bill of Rights* (1960), which is still in force, the Charter represents an essential constitutional document. The Charter is part of the "supreme law of Canada" by virtue of its inclusion in the Constitution of Canada. Also, as part of the Constitution, any law that is not consistent with the provisions of the Charter is, "to the extent of the inconsistency, of no force or effect." This simply means that laws that are found to be contrary to the Charter cannot be applied. In effect, a law that offends the Charter loses its legal power. For example, a law dealing with warrantless searches, if found to be contrary to the Charter, will be struck down by the courts and rendered ineffective. The Charter is, then, a special piece of legislation; it is a constitutional instrument whose interpretation follows a different path than ordinary statutes (McDonald, 1982).

In considering the origins of the Charter, it has been noted that the late Pierre Elliott Trudeau played a pivotal role (Manning, 1983):

> The Canadian Charter of Rights and Freedoms was really the brain-child of the then Minister of Justice and Attorney General of Canada Pierre Elliott Trudeau. In 1955, in a proposal to the Quebec Royal Commission of Inquiry on Constitutional Problems he stated that Quebec should declare itself ready to accept the incorporation of human rights into the Constitution of Canada and a precise plan should be mapped out for patriating the Canadian Constitution. (p. 1)

The Charter was introduced in order to provide a solid basis for making determinations about individual rights in Canada. It should be seen within an extensive, international context during the 1970s and 1980s where the topic of human rights was significant on the public agenda. Manning (1983) notes the following:

> . . . our Charter is substantially similar in many ways not only to the American Bill of Rights, but to the United Nations Declaration of Human Rights adopted in 1948, the European Convention on Human Rights and Fundamental Freedoms which came into force in 1953, and the International Covenant on Civil and Political Rights of 1966, to which Canada became a signatory in 1976. Those provisions have been examined by judges of all levels of courts in the United States, by the European Commission of Human Rights and by the European Court of Human

Rights and have been applied to fact situations which will arise in Canada. Their decisions will be referred to in the hope that their experience will assist the Canadian judiciary when they come to deal with the application of the Charter. (p. 3)

It falls to our Canadian judiciary to determine the extent to which the Charter will afford protection for individuals, as well as to the politicians we elect to represent us and who make appointments to our courts (Manning, 1983, at p. 2). The Charter serves as a symbol of our commitment to human rights, and as a mechanism for actually demonstrating that commitment.

The introduction of the Charter has certainly increased the prominence of the judiciary, particularly within the Supreme Court of Canada. It has served to focus considerable attention on the nine justices who constitute the country's highest court (Manning, 1983):

> For the first time the explicit guarantees of civil liberties in a constitutional document call upon the judiciary to be the *ultimate expounder* of the meaning and the range of our liberties. It is to the judiciary that the public must now look to determine the scope and extent of their fundamental rights and freedoms. (p. 21) [emphasis added]

As we can see, it ultimately rests with the Supreme Court of Canada to decide if police powers of detention, arrest, and search and seizure are consistent with the principles of justice within a Canadian context.

SIGNIFICANCE OF THE CHARTER IN CANADIAN CONSTITUTIONAL LAW

The Charter plays a leading role in the context of Canadian constitutional law. Since its passage in 1982, the Charter has become a critical element in the life and work of the country's highest court, the Supreme Court of Canada. Constitutional law may be seen as the collection and interpretation of laws that guide the relationship between individuals and the state, and between various parts of the state. As one author (Milne, 1982) observes:

> The constitution in a federal state forms a master power grid from which all the governmental players get their authority. (p. 21)

Because the Constitution is the supreme law in Canada, it is used to determine the validity of all other laws and statutes, federal and provincial.

The Charter, incorporating important issues of rights and freedoms, may be employed to resolve central constitutional research problems (Castel and Latchman, 1996). Therefore, anyone interested in learning about police powers and their relationship with individual rights and freedoms will need to know how to study the Charter and its application on an ongoing basis. There are several tools available that will assist in this regard (Castel and Latchman, 1996, pp. 78–89) and students are well-advised to become familiar with these aids to legal research.

IMPACT OF THE CHARTER ON POLICING IN CANADA

Because police officers in Canada are agents of the state, with a considerable degree of discretion in the execution of their duties, they have a great deal to do with the liberty of individual citizens. No other public organization can directly impact the rights and freedoms of individual citizens to the extent of the police. Police officers have a special legal responsibility that allows them to severely limit our freedoms, and the law extends to them the authority to use whatever means necessary, including lethal force, to constrain our liberties under certain circumstances.

The Charter has had a profound impact on police powers in Canada and continues to influence the way police officers function across the country. It is safe to suggest that no single piece of legislation, with the possible exception of the *Criminal Code of Canada*, has as much power to direct individual officers in their day-to-day functions as the Charter.

Credit: Mike Weaver, Media Relations, Kingston Police

The Charter has an impact on every police officer in Canada.

SPECIFIC SECTIONS OF THE CHARTER

While the Charter, in its entirety, is vitally important to a complete understanding of Canadian constitutional law, there are essentially seven sections of that document that are particularly relevant to police officers in learning about the nature and scope of their powers. These sections warrant special attention in any program designed for learning about arrest, search and seizure, or other police activities aimed at the enforcement of Canadian criminal law. Each section dealt with below has received extensive treatment by the courts, legal scholars, and others concerned with understanding the impact of the Charter on policing. Each of those sections will be outlined below; however, there will be a more detailed discussion of the pertinent sections in the chapters that follow. It is useful to become familiar with the wording and focus of each of the highlighted sections in advance of our more extensive examination.

The Charter is a special piece of legislation, and it is important to fully understand its elements, including the limitations to which it is subject. Section 1 reads as follows:

> The *Canadian Charter of Rights and Freedoms* guarantees the rights and freedoms set out in it subject only to such reasonable limits prescribed by law as can be demonstrably justified in a free and democratic society.

This raises specific questions about the nature of a "free and democratic society" which are beyond the immediate scope of this textbook. However, it is clear that any allowable limits must be reasonable and there must be some demonstration of the justification for those limits. The test of any limits on rights and freedoms will be derived from a wide range of sources, including public opinion, community standards, jurisprudence, sociology, political theory, ethics, and many other sources of insight relevant to matters pertaining to a so-called free and democratic society. Section 1 of the Charter deals with the concept of the reasonable limits that may be placed on the protections afforded rights and freedoms in Canada. This section outlines the parameters that can be applied within "a free and democratic society" and, of course, opens the question as to what constitutes the kinds of reasonable limits that might be imposed in such a society.

Section 7 of the Charter states:

> Everyone has the right to life, liberty and security of the person and the right not to be deprived thereof except in accordance with the principles of fundamental justice.

This section presents a relatively broad conceptual framework that allows for considerable debate over the precise meaning of the terms used. The individual terms "life," "liberty," and "security of the person" are certainly open to further examination, especially when one considers the impact of modern technology, medical ethics, or, the influence of global issues like environmentalism, international terrorism, and other topics that might influence our personal security. Also, it is certain that a precise explanation of "the principles of fundamental justice" is not going to be as readily available as a clear proof of the Pythagorean theorem. However, the fact that these concepts are difficult to articulate does not in any way diminish their importance for police officers learning about the Charter. The context in which Canadian police officers function is defined by this document and each of the sections holds clues to the interpretations the courts will provide on their specific, day-to-day activities. Also, the justices themselves go to great lengths to articulate the concepts contained in the Charter by utilizing the most compelling research and supporting evidence to clarify their judgments on issues before the court.

Section 7 tells us that the Charter takes individual rights seriously and that those rights may only be infringed upon in ways that can be justified in a society that places high value on the concept of justice. It tells us that the protection of life, liberty, and security of the person are things that matter deeply to us, as citizens and human beings, and that it is only under exceptional circumstances that these rights may be in any way limited. It further tells us that any limitation must be seen as a "deprivation" and will only be seen as acceptable if it is consistent with notions of fundamental justice that may be said to transcend the practicalities of everyday life.

Section 8 of the Charter states:

> Everyone has the right to be secure against unreasonable search or seizure.

This section bears some comparison to the Fourteenth Amendment of the United States Bill of Rights, which reads as follows:

> The right of the people to be secure in their persons, houses, papers and effects, against unreasonable searches and seizures, shall not be violated, and no Warrants shall issue, but upon probable cause, supported by Oath or affirmation, and particularly describing the place to be searched and the persons or things to be seized.

There has been a tradition of looking to American jurisprudence for assistance in making determinations about the application of the Charter. The concept of "persuasive authority" should be understood in this context. For example, as Dambrot (1982) explains:

> While it might be desirable to avoid excessive reliance on American cases, it seems inevitable that they will assume an important role in the development of our law, both because of the extensive experience in the United States with an exclusionary rule (while we have virtually none), and because the language of s. 8 seems to have been derived from the American Constitution. (p. 97)

On a specific point, when looking at a search and seizure that has been approved by an appropriate judicial authority, there is little expectation that it would be found to be unreasonable. Be that as it may, such judicial approval does not exempt a warrant from review (McDonald, 1989):

> A search carried out pursuant to a search warrant obtained without the information in support of the application being in sufficient detail is unreasonable. While in many such cases even without the Charter the search warrant would be held to be invalid, now s.8 of the Charter dictates that the requirements of s.443 of the Code be interpreted in light of s.8. These standards must be considered the minimal standards to be met before a warrant will be properly issued. The right to intrusion must be determined in advance by an impartial judicial officer, and not left to the police. (p. 251)

Section 8 has therefore expanded the interest of the Bar in pursuing the "manner" in which a warrant has been obtained. Every aspect or element of the process may be scrutinized to ensure that the requirements of the Charter, and by extension, the rights and freedoms of anyone subject to the search and seizure, have been observed.

It is important to understand the scope of section 8, particularly as it pertains to the guarantee against unreasonable search and seizure. To gain this understanding it is helpful to compare that section to the U.S. Bill of Rights (McDonald, 1982):

> There is one feature of the Canadian Charter which differs from the United States Bill of Rights. The latter is directed against the acts of agents of governments; consequently, an act of a private individual who is in no sense the instrumentality of government is held not to be within the constitutional proscription against unreasonable searches and seizures. Section 8 of the Canadian Charter, on the other hand, broadly guarantees "the right to be secure against unreasonable search or seizure", and neither the section on its face nor the context of the section limits the right to security against unreasonable search or seizure to situations in which the search or seizure is by an agent of government . . . there may be circumstances when evidence has been seized by a private person, the evidence is tendered in evidence for the Crown in a prosecution or for a party in a civil case, and the adversary will object on constitutional grounds to the admission of the evidence. (pp. 44–45)

The appearance of section 8 of the Charter is really something quite new in Canadian constitutional law and it therefore was anticipated as a method for more systematically controlling the application of police powers. The importance of such a constitutional innovation should be understood and emphasized (Manning, 1983, at p. 274):

> Section 8 of the Charter of Rights creates a right which was, in the terms used, hitherto unknown and which, when coupled with the broad discretionary remedy of section 24(1) and the exclusionary rule in section 24(2), will have a substantial impact on our jurisprudence. No similar provision existed in the Bill of Rights and while our law has always recognized a method of reviewing actions by police officers in searching and seizing, the control of these acts when found to be unlawful or illegal or even unreasonable has been minimal to say the least. This section was included to be used as a means of controlling not only the issuance of search warrants . . . but the use of those warrants. The section will also be used to monitor all searches and seizures in relation to regulatory offences both federal and provincial in areas such as customs, excise, wild life protection, environmental protection and health.

The scope of section 8 was intended to be significantly broader than previous constitutional legislation. In looking for a source of this section, it has been suggested that there was ample evidence provided by the McDonald Commission into certain activities of the RCMP. The relevance of contemporary Canadian events on the construction of the Charter is not to be overlooked, or diminished (Manning, 1982, at pp. 278–79):

> Section 8 did not come into being as a protection against the type of abuses that gave rise to the American Fourth Amendment. We must look to our own history in order to gain some insight into the passage of that section. There is no doubt that the framers of section 8 of the Charter had in mind the inquiry into R.C.M.P. activities in the MacDonald [sic] Commission hearings. In the Royal Commission of Inquiry concerning certain activities of the Royal Canadian Mounted

Police . . . there was revealed activities by the police acting without a warrant or acting with a warrant but without any jurisdiction in the justice issuing a warrant. These, undoubtedly, were the kinds of activities that were sought to be curtailed by section 8 and it would be a mistake to ignore the facts revealed in that Report or the public condemnation of those activities.

Generally speaking, public inquiries, government law reform studies, and Royal Commissions often have a lasting and meaningful impact on constitutional law, as well as on the judiciary when they are given the task of reviewing Charter cases. The importance of section 8 of the Charter was certainly predicted early in its existence. Its impact on police powers has been considerable as Manning assured us it would be (1983, at p. 322):

Section 8 promises to be one of the more regularly litigated provisions in the Charter and future decisions should reveal how ready the Canadian judiciary is to follow American authority at a time when there appears to be a retreat from the "blanket exclusionary rule" in the United States.

Section 9 of the Charter states:

Everyone has the right not to be arbitrarily detained or imprisoned.

This section will be discussed in detail in Chapter 3, below.
Section 10 of the Charter reads:

Everyone has the right on arrest or detention
(a) to be informed promptly of the reasons therefor;
(b) to retain and instruct counsel without delay and to be informed of that right; and
(c) to have the validity of the detention determined by way of *habeas corpus* and to be released if the detention is not lawful.

McDonald (1982) is helpful when he makes a comparison between the language of this section and the European Convention on Human Rights:

Motorcyle gangs in Canada present a special challenge to police services

Credit: Michael Lea/Kingston Whig-Standard

The clause in the European Convention alerts us to the possibility that in Canada the word "informed" should be interpreted in a manner which is subjective to the person arrested—in other words, that the person arrested must be "informed" of the reasons for his arrest or detention in a language he understands. Such an argument, if successful, would, from the practical point of view, impose a duty on the police to have interpreters available on a reasonably steady basis, particularly in localities where there are significant numbers of people, such as native people and immigrants, who do not understand the English (or French) language. (p. 77)

There is a special onus on a police officer to ensure that an accused has access to legal counsel. The onus is on the officer to provide access to counsel, but a breach does not *necessarily* discount evidence (McDonald, 1989, at p. 346):

The provision in the Canadian Bill of Rights has not received extensive judicial interpretation. The leading case is Hogan v. R., [1975] 2 S.C.R. 574 where the majority of the Supreme Court of Canada held that a policeman's refusal to allow the accused access to his counsel, whom he had already retained and who was to his knowledge at the police station where the accused was, violated his right to instruct counsel under section 2(c)(ii) of the Canadian Bill of Rights; however this did not result in the evidence of the result of the breath test he then took being inadmissible. The majority held that common law prevails, that relevant evidence is admissible, no matter how it is obtained, and that the Canadian Bill of Rights has no impact on that rule.

The Charter adds the requirement that the police are to inform persons of their right to retain and instruct counsel, without delay. McDonald (1982) points out that Canadian law was guided by the important *Miranda* decision in the United States:

Thus the Canadian Charter falls in line with the position in Gideon v. Wainwright [(1964), 378 U.S. 478], in the United States Supreme Court, which held that the failure of a state court to advise a criminal defendant of his right to be represented by counsel is a violation of the Fourteenth Amendment. In the famous case of Miranda v. Arizona, [(1966), 384 U.S. 436 at 471] this was held to apply also to the stages of preliminary examination and, still earlier, to police interrogation of an individual held in custody. (p.79)

Section 11 of the Charter outlines the rights of an accused. There are nine elements defined in that section, which reads as follows:

Any person charged with an offence has the right

(a) to be informed without unreasonable delay of the specific offence;
(b) to be tried within a reasonable time;
(c) not to be compelled to be a witness in proceedings against that person in respect of the offence;
(d) to be presumed innocent until proven guilty according to law in a fair and public hearing by an independent and impartial tribunal;
(e) not to be denied reasonable bail without just cause;
(f) except in the case of an offence under military law tried before a military tribunal, to the benefit of trial by jury where the maximum punishment for the offence is imprisonment for five years or a more severe punishment;

(g) not to be found guilty on account of any act or omission unless, at the time of the act or omission, it constituted an offence under Canadian or international law or was criminal according to the general principles of law recognized by the community of nations;

(h) if finally acquitted of the offence, not to be tried for it again and, if finally found guilty and punished for the offence, not to be tried or punished for it again; and

(i) if found guilty of the offence and if punishment for the offence has been varied between the time of commission and the time of sentencing, to the benefit of the lesser punishment.

Section 24 of the Charter states:

(1) Anyone whose rights and freedoms, as guaranteed by this Charter, have been infringed or denied may apply to a court of competent jurisdiction to obtain such remedy as the court considers appropriate and just in the circumstances.

(2) Where, in proceedings under subsection (1), a court concludes that evidence was obtained in a manner that infringed or denied any rights or freedoms guaranteed by this Charter, the evidence shall be excluded if it is established that, having regard to all the circumstances, the admission of it in the proceedings would bring the administration of justice into disrepute.

In the event that a person's rights under the Charter have been violated, there is provision for the exclusion of evidence. Much of the groundwork for the exclusionary rule can be found in the Law Reform Commission of Canada's study *The exclusion of illegally obtained evidence: a study paper* (1974):

> Canadian law has followed English law: The illegality of the means used to obtain evidence generally has no bearing upon its admissibility. If, for example, a person's home is illegally searched—without a search warrant or reasonable and probable cause for a search—the person may sue the police for damages incurred, complain or demand disciplinary action or the laying of criminal charges. But, the evidence uncovered during this search together with all evidence derived from it is admissible.

Of course there is an opposing view that inclines to fairly blunt treatment for the criminal and resists the inclination to place too many hurdles in the way of the police when executing their duties and responsibilities. For example, see the case of *R. v. Honan* (1912), 26 O.L.R. 484, 20 C.C.C. 10 at 16, 6 D.L.R. 276 (C.A.) where Mr. Justice Meredith says:

> The criminal who wheels the "jimmy" or the bludgeon, or uses any other criminally unlawful means or methods, has no right to insist upon being met by the law only when in kid gloves or satin slippers; it is still quite permissible to "set a thief to catch a thief" . . .

CONCLUSION

We have considered the central importance of the Charter as a legal document that will guide police powers in the critical areas of arrest, detention, and search and seizure. The origins of the Charter as part of the patriated Constitution have been reviewed and we have briefly considered the seven key sections of the Charter that have a substantial impact on the manner in which police conduct their day-to-day activities.

By considering the pivotal role of the Canadian judiciary and the case law that continues to grow around the key sections of the Charter, readers should be able to refine their understanding and knowledge of these provisions. They should also attempt to keep abreast of changes to the law as they pertain to issues of police power and its constitutional limits.

We will delve more deeply into the pertinent sections of the Charter in the following two chapters; however, it remains important that the reader comprehend the basis for this constitutional mechanism for gauging and guiding police powers in Canada.

QUESTIONS FOR CONSIDERATION & DISCUSSION

1. Does the Charter enhance or hinder the ability of the police to function?
2. Is it appropriate for the Supreme Court of Canada to take such a critical role in the determination of Charter issues?
3. How can Canadian police organizations ensure that they are fully aware of the impact of Charter decisions?
4. What, if any, are the differences in the protections provided for individual rights and freedoms between Canada and the United States?
5. Has the introduction of the Charter improved individual rights and freedoms in Canada?

CASES CITED

Gideon v. Wainwright, 378 U.S. 478 (1964)

Hogan v. R., [1975] 2 S.C.R. 574

Miranda v. Arizona, 384 U.S. 436 (1966)

R. v. Honan (1912), 26 O.L.R. 484, 20 C.C.C. 10 at 16, 6 D.L.R. 276 (C.A.)

R. v. Hynds (1982), 8 W.C.B. 182 (Alta. Q.B.)

U.S. v. Mendenhall, 446 U.S. 544 (1980)

REFERENCES

Castel, Jacqueline R. and Omeela K. Latchman (1996). *The practical guide to Canadian legal research.* 2nd ed. Toronto: Carswell.

Commission of Inquiry Concerning Certain Activities of the Royal Canadian Mounted Police (1981). *Report of the Commission of Inquiry Concerning Certain Activities of the Royal Canadian Mounted Police.* Ottawa: Minister of Supply and Services Canada.

Dambrot, Michael R. (1982). "Section 8 of the Canadian Charter of Rights and Freedoms." In 26 C.R. (3d) 97.

Law Reform Commission of Canada (1974). *The exclusion of illegally obtained evidence: a study paper*. Ottawa: The Commission.

Manning, Morris (1983). *Rights, freedoms and the courts: a practical analysis of the Constitution Act, 1982*. Toronto: Emond-Montgomery.

McDonald, David C. (1982). *Legal rights in the Canadian Charter of Rights and Freedoms: a manual of issues and sources*. Toronto: Carswell.

McDonald, David C. (1989). *Legal rights in the Canadian Charter of Rights and Freedoms: a manual of issues and sources*. 2nd ed. Toronto: Carswell.

Milne, David (1982). *The new Canadian constitution*. Toronto: James Lorimer & Company.

RELATED ACTIVITIES

- Read Charter cases in various series of Canadian law reports (e.g., *Canadian Criminal Cases, Criminal Reports*) or other works that focus attention on this area.
- Scan local and/or national newspapers and Web sites for items on the cases relating to the *Canadian Charter of Rights and Freedoms*, or, items that deal with policing and constitutional rights. Examples of relevant topics include search and seizure, RIDE programs, the stopping of outlaw motorcycle gang members, and police arrests.
- Examine the Supreme Court of Canada's Web site at **www.scc-csc.gc.ca** and study individual decisions pertaining to the specific sections of the Charter examined in this chapter.

WEBLINKS

 http://www.scc-csc.gc.ca/ An introduction to the Supreme Court of Canada, including its role, cases, judgements, and FAQs.

 http://www.extension.ualberta.ca/legalfaqs/nat/char.htm Go to this site for questions and answers regarding the Charter and criminal law.

 http://www.pch.gc.ca/ddp-hrd/english/charter/contents.htm This page provides an overview of the Charter, the full text, and commentary.

CHAPTER TWO

POLICE POWERS OF

ARREST AND RELEASE

LEARNING OBJECTIVES

1. List sections of the *Criminal Code of Canada* that pertain to arrest, with or without a warrant.
2. Provide a definition for "arrest" in Canadian law.
3. Provide a definition for "detention" in Canadian law.
4. Distinguish between an arrest with and without a warrant.
5. List the general powers of arrest for a peace officer in Canada.
6. List the powers of arrest of anyone in Canada.

INTRODUCTION

The importance of the police decision to arrest cannot be overstated. It marks a significant event at the front end of the criminal justice system. Once an arrest has occurred, the person taken into custody becomes subject to the judicial process to a greater, or lesser, extent with potentially dramatic repercussions with regard to that person's liberty.

This chapter will provide more detailed background on the topic of arrest, one of the significant police powers exercised in Canada. We will explore the Charter provisions that relate to detention and arrest and consider several important criminal cases that have helped to define these key terms. This chapter will proceed from the general to the specific and will cite relevant sections of the *Criminal Code of Canada* as it pertains to police powers of arrest.

It is worth repeating that readers should be alert to the potential for change in the context of Canadian criminal law. Not only are statutes amended through the federal Parliament, but there is ongoing action on this front through court decisions, especially those rendered by the Supreme Court of Canada. The presentation which follows reflects the state of the law at the time of publication, and students will need to ensure that modifications have not occurred in these areas.

ARREST

Because the topic of arrest is relatively detailed and complex, the following outline of this chapter's contents is provided by way of summary:

1. Preamble
2. What constitutes an arrest?
3. Detention for investigation
4. The *Canadian Charter of Rights and Freedoms*
   Rights upon arrest or detention
   Meaning of "on arrest or detention"
   The nature and meaning of the right to counsel
   Protection against arbitrary detention
   Consequences of a breach of the Charter
5. General powers of arrest without a warrant
6. Specific powers of arrest
7. Limitations on powers of arrest
8. Procedure to be followed where the peace officer does not arrest
9. Duties after arrest
10. Laying of an information
11. Arrest with a warrant
12. When may a warrant be issued?
13. Carrying the warrant
14. Execution of the warrant where the accused is in another province
15. Summons
16. Questioning a suspect
17. Arrest of wrong person or on faulty grounds
18. Searching an arrested person
19. Arresting a young offender
20. Trespass in making an arrest
21. Assisting a peace officer
22. Refusal of citizen to identify self
23. Consequences of an illegal arrest
24. Arrest under provincial statutes and by-laws
    Ontario *Highway Traffic Act*
    Ontario *Liquor Licence Act*
    Ontario *Game and Fish Act*
    Municipal by-laws

PREAMBLE

Before an accused person may be prosecuted, that person must first be brought forward to appear in court. This may be accomplished in one of the following two ways:

1. A summons or appearance notice is issued; or
2. An arrest with or without a warrant is accomplished.

Prior to 1972, peace officers exercised a considerable degree of discretion as to whether or not to carry out an arrest. As a result of the work of the 1968 Ouimet Commission, which examined criminal justice and corrections in Canada, it was determined that peace officers did not always use their discretion wisely and that specific restrictions should be placed on their discretion. Accordingly, the *Bail Reform Act* was passed. The *Bail Reform Act* requires that peace officers consider the public interest, and whether or not it can best be served with or without an arrest. Section 495(2) of the *Criminal Code* specifically directs a peace officer *not* to arrest a person without a warrant for:

- An indictable offence which is within the absolute jurisdiction of a provincial court judge or a judge of the Nunavut Court of Justice;
- An offence for which a person may be prosecuted either on an indictment or summary conviction;
- A summary conviction offence; or
- An offence for which the "public interest" may be satisfied without arresting the person and where the peace officer does not have reasonable grounds to believe that the accused will not appear for his or her trial.

Because the term "public interest" is not defined in the *Criminal Code* it has been seen to be vague and imprecise. In the case of *Morales* (1992), 77 C.C.C. (3d) 91, 17 C.R. (4th) 742 (S.C.C.) it was found that detention would be unconstitutional because the courts could define any circumstance as appropriate for pre-trial detention. But, section 495(2) of the *Criminal Code* identifies three specific factors that the police may take into account to determine if an accused should be arrested in the public interest. Those factors are as follows:

1. The need to establish the person's identity;
2. The need to secure or preserve evidence relating to the offence; and
3. The need to prevent the continuation or repetition of the offence or the commission of another offence.

Therefore, pre-trial detention is restricted to these very precise factors. The police officer is prevented from making an arrest unless there are reasonable grounds to believe that the public interest will not be satisfied by releasing the accused, or that the accused will not show up for trial.

Prior to the introduction of the Charter, if a police officer failed to comply with the section 495(2) limitation of the *Criminal Code*, the arrest was generally still considered lawful and did not impact on the validity of the prosecution. Recently, however, Canadian courts have been critical of the failure of the police to respect the language and spirit of this section. It has been seen to amount to an arbitrary detention of an accused, which is contrary to section 9 of the Charter. Therefore, as a matter of general policy, it is *not* appropriate for the police to arrest anyone who commits a specific kind of offence where the "public interest" is not affected. This is the finding in the case of *Labine* (1987), 49 M.V.R. 24 (B.C. Co. Ct.).

Accordingly, peace officers must look at each case on an individual basis and use their best judgment along with the information available at the time. According to section 495(2) of the *Criminal Code*, unless an officer has reasonable grounds to believe either that it is not in the "public interest" to release the suspect or that the suspect will fail to show up for trial, the officer's duty is *not* to arrest the suspect.

WHAT CONSTITUTES AN ARREST?

To arrest means to stop, stay, or restrain the liberty of someone. A police officer may stop someone on the street to question that person. Certain provincial statutes (like the *Highway Traffic Act* in Ontario) allow a peace officer to stop a motorist on the highway for certain specified purposes. Also, the *Criminal Code* gives the authority to a peace officer to search a suspect under certain circumstances (e.g., for prohibited or restricted weapons, firearms, or ammunition). Subsection 11(5) of the *Controlled Drugs and Substances Act* authorizes the search of someone for a narcotic during the execution of a search warrant. The *Food and Drugs Act* also has similar provisions.

On the question of whether or not a person should consider themselves "seized" or detained by the police, Manning (1983, at p. 288) provides the following assistance:

> A test that may be applied in determining whether a person has been "seized" is whether in view "of all the circumstances surrounding the incident, a reasonable person would have believed that he was not free to leave" [U.S. v. Mendenhall, 446 U.S. 544 (1980)]. A determination of whether such freedom of choice exists involves "a refined judgment" especially when no force, physical restraint or blatant show of authority is involved . . . This judgment may involve looking at matters as nebulous as the tone of an officer's question.

Credit: Mike Weaver, Media Relations, Kingston Police

A suspect is taken into custody by the police.

There is a symbolic element involved in arrest that includes physical contact between the police officer and the person being arrested. The touching of a person is sufficient to indicate that an arrest has taken place. Canadian courts have considered the issue (Manning, 1983, at p. 289):

> In R. v. Whitfield, [1970] S.C.R. 46, the Supreme Court of Canada held that there was no room for what seemed to be a new subdivision of "arrest" into "custodial" arrest and "symbolical" or "technical" arrest. It held that an arrest was effected by an actual touching or seizure of a person's body with a view to his detention or by words of arrest if the person arrested submitted to the process and went with the arresting officer.

It is important to recognize that any stopping of a person is, in fact, a restraint of that person's liberty. In Canadian law, however, an arrest does not take place until the officer has actually touched the person. A mere statement to the effect that someone is under arrest is not sufficient under the law. As Salhany (1997) indicates:

> An arrest may occur where words are used which clearly indicate to the person that if he attempts to leave, he will be physically restrained and the person does in fact submit to his loss of liberty. (p.26)

Police officers frequently wish to question people, without arresting them. Canadian law recognizes that police officers must investigate crimes as part of their sworn duty, and this requires that they have a right to question citizens, perhaps even insisting that they answer their questions. However, against this right is a competing right for citizens to refuse to answer questions. If a police officer asks someone to accompany him or her to the police station, and that person goes willingly, no arrest has occurred. If the person refuses and detained or forced to go to the station, an arrest has occurred. If the person is led to believe that he or she will be taken into custody upon refusing to accompany an officer, then an arrest has occurred. This matter is addressed in the cases of *Koechlin v. Waugh and Hamilton* (1957), 118 C.C.C. 24 (Ont. C.A.) and *Cluett* (1983), 3 C.C.C. (3d) 333 (N.S.C.A.) and in the *O'Donnell* decision.

DETENTION FOR INVESTIGATION

In the case of *R. v. Chromiak* (1979), 49 C.C.C. (2d) 257 (S.C.C.), the appellant had refused to provide a breath sample for the purposes of a roadside screening test to determine blood alcohol content. At issue in this case was whether denial of right to counsel constituted a reasonable excuse for refusing, and whether or not the accused was "detained."

Speaking on behalf of the court, Justice Ritchie made the following observation:

> . . . I have concluded that the appellant was not a person who had, while "arrested or detained" been deprived of the right to "retain and instruct counsel without delay" and I am unable to find any reasonable excuse for his failure to comply with the demand made to him by the peace officer. (p. 263)

The case of *Dedman* (1981), 59 C.C.C. (2d) 97 (Ont. C.A.) at 109, affirmed 20 C.C.C. (3d) 482 (Ont. C.A.) made it clear that a police officer did not have the right to detain a suspect for questioning without proof on the basis of reasonable probability. This understanding was qualified by Mr. Justice Doherty in the case of *Simpson* (1993), 79 C.C.C. (3d) 482 (Ont. C.A.). Here it was determined that the *Dedman* case did not prevent a police officer from detaining a suspect during the course of an investigation, even though the officer's reasons did not meet the standards for arrest. Mr. Justice Doherty found that the officer's power to detain was derived from the *common law* and still exists. The standard for an investigation should be one of "reasonable suspicion" as distinguished from reasonable cause. The officer must be able to point to specific facts that can be considered objectively in weighing whether or not the officer had reasonable cause to believe that the suspect was involved criminally in the activities under investigation. For example, in the *Simpson* case, evidence of cocaine was held to be inadmissible because the officer was not able to establish that his opinion was based on factors that could be articulated to the satisfaction of the court. The detention in this case was unreasonable because Constable Wilkin clearly acknowledged that his decision to stop the car was completely unrelated to the operation of a motor vehicle. Furthermore, the officer was not relying on any specific legal authority when making the stop. Therefore, with no statutory authority, and no common law authority, Constable Wilkin's conduct was found to be unlawful.

The case of *R. v. Therens* (1985), 18 C.C.C. (3d) 481 (S.C.C.) deals directly with the actual meaning of "detention." In this case, the court determined that Mr. Therens was under detention, and, as a result, his section 10(b) rights under the Charter had been infringed. Specifically, in this case, the demand under section 235(1) of the *Criminal Code* requiring the accused to accompany a police officer to a police station to submit to a breathalyzer test amounted to a detention under section 10 of the Charter.

As part of his dissenting opinion, Justice Le Dain noted the following:

> The purpose of s. 10 of the Charter is to ensure that in certain situations a person is made aware of the right to counsel and is permitted to retain and instruct counsel without delay. The situations specified by s. 10—arrest and detention—are obviously not the only ones in which a person may reasonably require the assistance of counsel, but that are situations in which the restraint of liberty might otherwise effectively prevent access to counsel or induce a person to assume that he or she is unable to retain and instruct counsel. In its use of the word "detention", s. 10 of the Charter is directed to a restraint of liberty other than arrest in which a person may reasonably require the assistance of counsel but might be prevented or impeded from retaining and instructing counsel without delay but for the constitutional guarantee. (pp. 503–504)

Of particular relevance to those seeking to understand the nature and extent of police powers in Canada are Le Dain's additional observations on the level of knowledge of the average citizen with respect to police authority:

> Most citizens are not aware of the precise legal limits of police authority. Rather than risk the application of physical force or prosecution for willful obstruction, the reasonable person is likely to err on the side of caution, assume lawful authority and comply with the demand. The element of psychological compulsion, in the form of a reasonable perception of suspension of freedom of choice, is enough to make the restraint of liberty involuntary.
>
> Detention may be effected without the application or threat of application of physical restraint if the person concerned submits or acquiesces in the deprivation of liberty and reasonably believes that the choice to do otherwise does not exist. (p. 505)

THE *CANADIAN CHARTER OF RIGHTS AND FREEDOMS*

As we examined in Chapter 1, the Charter became law in Canada on April 17, 1982. It entrenches certain rights and freedoms that have already existed. Some of these rights and freedoms have been recognized under common law for centuries; others were recognized in the *Canadian Bill of Rights* passed in 1960, S.C. 1960, c. 44.

The Charter gave courts the power to impose sanctions on the police, or government agencies, when they have been found to breach the rights of citizens. Specifically, section 24 of the Charter authorized the courts to impose sanctions, or remedies, when a person's rights have been breached.

Section 10 of the Charter provides as follows:

Everyone has the right on arrest or detention

(a) to be informed promptly of the reasons therefor;
(b) to retain and instruct counsel without delay and to be informed of that right; and
(c) to have the validity of the detention determined by way of *habeas corpus* and to be re-leased if the detention is not lawful.

RIGHTS UPON ARREST OR DETENTION

Basically, there are three rights included in this section:

1. To be informed promptly of the reasons for arrest or detention;
2. To be informed about the right to counsel; and
3. To retain and instruct counsel.

As a useful point of reference for all aspects of Canadian criminal law, Salhany (1997) offers the following instruction:

> It is a general rule of statutory interpretation that whenever Parliament passes a law, each word used in the section is intended to have a specific and distinct meaning. (p. 29)

MEANING OF "ON ARREST OR DETENTION"

The *Criminal Code* supports the notion of making a distinction between "detention" and "arrest." The word "detain" is used in sections 28(2)(b), 30, 148(a), and 281 while the word "arrest" is used in various other parts of the *Criminal Code*. Certainly, detention implies something less than an arrest. The case of *Chromiak* (1979), 49 C.C.C. (2d) 257 (S.C.C.) offered a view of the Supreme Court of Canada prior to the enactment of the Charter. This view was, however, altered by the Supreme Court in the *Therens* case (1985), 18 C.C.C. (3d) 481 (S.C.C.). In *Therens*, the court noted that most people are not familiar with the limits on police authority, and therefore they are inclined to err on the side of caution in their dealings with the police, assuming that officers have a lawful authority to make certain demands. As a result, detention may be effected without the application or the threat of application of physical restraint because a person allows herself or himself to be deprived of liberty based on the reasonable belief that she or he has no choice.

There are certain limitations set by the Charter based on section 1, which addresses the "reasonable limits prescribed by law as can be demonstrably justified in a free and democratic society." For example, a right to counsel should be limited in a case involving a roadside screening device (under section 254(2) of the *Criminal Code*) because this is an investigative tool that contributes not only to detection, but also to deterrence.

It must be understood that there are occasions when the police will interact with individuals in the community for reasons other than arrest or detention. Salhany (1997) is helpful on this point:

> On the other hand, the courts have recognized that not every contact between a police officer and a citizen will necessarily cause a psychological restraint which amounts to a detention. (p. 32)

What does and does not constitute a detention has been illustrated in several court decisions, which are discussed in the following paragraphs.

The case of *Esposito* (1986), 24 C.C.C. (3d) 88 (Ont. C.A.) considers the circumstances under which someone was questioned in his home by the police about stolen credit cards. Following questioning, Mr. Esposito was arrested and advised of his rights. The court held that there was no evidence that he was subject to a demand or direction that would lead him to feel that he was compelled to comply. His right to counsel was, therefore, not breached. Also, in the case of *Bazinet* (1986), 25 C.C.C. (3d) 272 (Ont. C.A.), the court held that no detention had taken place because the officer testified that Mr. Bazinet was free to leave the police station anytime he requested.

The finding was different in the case of *Voss* (1989), 50 C.C.C. (3d) 58 (Ont. C.A.). In this instance, a police officer asked Mr. Voss to accompany him to the police station to give a statement about his wife's death. When Voss was confronted with the results of an autopsy, he made statements that implicated him in his wife's death. While those statements were admitted by the trial judge, when the case was reviewed by the Ontario Court of Appeal it was found that Voss's statement took place under circumstances that would constitute detention; i.e., there was the necessary element of compulsion or coercion involved.

On appeal in the case of *Hawkins* (1993), 20 C.R. (4th) 524 (Nfld. C.A.), the Supreme Court of Canada held that there can be no detention without an actual sense of compulsion in the mind of the person being investigated.

THE NATURE AND MEANING OF THE RIGHT TO COUNSEL

Police need to be sensitive to the capacity and capabilities of the suspects when informing them of their right to counsel. Issues like ability to read, mental capacity, and understanding of the English language may all be taken into account in this regard.

In the case of *R. v. Vanstaceghem* (1987), 58 C.R. (3d) 121 (Ont. C.A.), the accused, an impaired driver, was not advised of his rights in French. He was seeking to have evidence against him excluded under section 24(2) of the Charter. (See box below.)

R. V. VANSTACEGHEM (1987), 58 C.R. (3D) 121 (ONT. C.A.)

Facts:

Mr. Vanstaceghem was found speeding at the Canadian Forces Base, Trenton. He was stopped by the Military Police (MP), who spoke to him in English, which he apparently understood. The accused responded to the officer in English and the MP provided a warning that Vanstaceghem said he understood.

Decision:

Lacourcière J.A. made the following observations:

> In my view the failure of the officer to inform the respondent of his rights in a meaningful way, that is to say, in French, indicates in the circumstances of this case a regrettable disregard for the respondent's constitutional rights. The admission into evidence of the breath sample emanating from the respondent following this failure would, in my view, tend to bring the administration of justice into disrepute. (p. 129)

In the case of *R. v. Manninen* (1987), 58 C.R. (3d) 97 (S.C.C.), the accused had been arrested for the robbery of a Mac's Milk in Toronto. Mr. Manninen was seeking an exclusion of evidence under section 24(2) of the Charter due to an infringement of section 10(b) of the Charter.

Mr. Justice Lamer speaking on behalf of the court noted that there had indeed been a violation of Manninen's right to counsel and accordingly allowed his appeal:

> The first point that must be made is that the violation of the respondent's right to counsel was very serious. The respondent clearly asserted his right to remain silent and to consult his lawyer. There was a telephone at hand. There was no urgency which would justify the immediate questioning or the denial of the opportunity to contact his lawyer. In effect, the police officers simply ignored the rights they had read to him and his assertion of the right to silence and the right to counsel. (p. 106)

In the case of *R. v. Tremblay* (1987), 60 C.R. (3d) 59 (S.C.C.), while the issue of right to counsel was central to the court, it became apparent that Mr. Tremblay's behaviour provided the basis for the court to dismiss his appeal, in spite of the fact that the police did not provide him with every reasonable opportunity to contact a lawyer. (See box below.)

R. V. TREMBLAY (1987), 60 C.R. (3D) 59 (S.C.C.)

Facts:

Accused was promptly informed of his right to counsel. He was given a telephone and called his wife.

Decision:

In providing the court's judgment, Justice Lamer offered the following:

> From the moment the accused was intercepted on the road to the moment he was asked to give the first sample of breath, his behaviour was violent, vulgar and obnoxious. A reading of the record and the findings of fact below satisfy me that, while the police, following the request for counsel, did not, as they must, afford the accused a reasonable opportunity to contact a lawyer through his wife before calling upon him to give a breath sample, their haste in the matter was provoked by the accused's behaviour. (p. 62)

In the case of *R. v. Clarkson* (1986), 25 C.C.C. (3d) 207 (S.C.C.), the issue of right to counsel is combined with the fact that the accused was found by the police in an intoxicated and highly emotional state. Clarkson was arrested on suspicion of murdering her husband. She was given a police caution and informed of her rights. Her aunt accompanied her and tried to prevent her from answering questions.

In this case, the concerns of the court were expressed by Justice Wilson:

> . . . the actions of the police in interrogating the intoxicated appellant seem clearly to have been aimed at extracting a confession which they feared they might not be able to get later when she sobered up and appreciated the need for counsel. In other words, this seems to be a clear case of deliberate exploitation by the police of the opportunity to violate the appellant's rights. (p. 220)

The fundamental nature of the right to counsel is the focus of the court's decision in the case of *R. v. Dubois* (1990), 54 C.C.C. (3d) 166 (Que. C.A.). (See box below.)

It is also necessary for the police to *re-inform* an accused person of the right to counsel whenever there has been fundamental or discrete change in the purpose of the investigation. This principle is discussed in *Black* (1989), 50 C.C.C. (3d) 1, 70 C.R. (3d) 97, [1989] 2 S.C.R. 138.

R. V. DUBOIS (1990), 54 C.C.C. (3D) 166 (QUE. C.A.)

Facts:

Mr. Dubois called the police to report his involvement in an accident. He stated that he had struck "someone or something" on the highway. Police arrived at the scene and found a victim lying alongside the road. Dubois was placed under arrest, advised of his constitutional rights and his right to "consult a lawyer." Dubois seemed in a state of shock. Accused was never advised that he could consult a lawyer "without delay."

Decision:

Speaking on behalf of the court, Fish J.A. noted the following:

> The right to counsel is a fundamental and indispensable characteristic of a free and democratic society. It is the sine qua non of due process and of a fair trial. But it is a hollow right indeed if those in acute need of its protection—persons under arrest or detention—are ignorant of their entitlement to the advice of a lawyer then and there. It is a hollow right as well if those in need, although informed of the right to counsel, are deprived of a reasonable opportunity to exercise it. And it is an unkept social promise if violation of the right to counsel is subject to no effective sanction and is condoned by the courts. (p. 175)

A valuable summary of the principles relating to detained persons, articulated by the court in this case, are reproduced below:

> The principles previously enunciated by the court, and still the law of the land, may now be summarized as follows:
>
> 1. Every person who is arrested or detained (whom I shall call a "detainee") must be informed without delay of his right to retain and instruct counsel without delay.
>
> 2. The detainee must be informed of this right in a manner which is comprehensible to him. The exact language of the Charter need not be used, so long as the detainee is clearly informed of every aspect of his right to counsel, including the right to retain and instruct counsel without delay.
>
> 3. In circumstances suggesting that the detainee does not understand the information communicated to him concerning his right to retain and instruct counsel without delay, a mere recitation of the right to counsel is insufficient. The police must take additional steps to ensure that the detainee is made aware of his s. 10(b) rights.
>
> 4. The police must refrain from seeking to elicit evidence from the detainee until he has been given a reasonable opportunity to retain and instruct

counsel of his choice, in private. Whether or not a reasonable opportunity has been provided depends on the circumstances of each case.

5. Once the police have informed the detainee, in a timely and comprehensible manner, of his right to retain and instruct counsel without delay, the detainee must exercise that right with reasonable diligence. If he fails to do so, the police are relieved of their duty to refrain from attempting to elicit evidence from him. By failing to act diligently, the detainee does not forfeit his right to counsel; he does, however, relieve the police of their duty to refrain from seeking to elicit evidence from him until he has exercised his s. 10(b) rights.

6. A detained person may waive his right to counsel, provided he appreciates the consequences of giving up the right. Any such waiver must be clear and unequivocal.

7. Where the detainee is properly informed of his s. 10(b) rights and there is nothing in the evidence to suggest that he did not understand those rights, compliance is presumed unless and until the detainee proves by a preponderance of evidence that he was denied a reasonable opportunity to ask for or to consult with counsel.

8. In the absence of evidence that the detainee failed to understand his right to counsel or was denied an opportunity to ask for counsel, no correlative duties are cast upon the police until the detainee, if he so chooses, has indicated his desire to exercise his right to counsel.

9. Because of the fundamental importance of the right to counsel in the administration of criminal justice, violation of s. 10(b) will generally result in exclusion under s. 24(2) of evidence closely connected with the violation. Exclusion, however, is not automatic.

10. In each case, the court must determine whether it is satisfied that admission of the evidence could—and not necessarily would—bring the administration of justice into disrepute. The test is not what the "public in general" or even a majority of the population might think. Rather, the judge himself or herself, examining all the circumstances carefully and impartially, must decide whether a reasonable person who understands the significance of the violation and the basic precepts of our system of justice, would consider admission or exclusion the greater evil.

11. Though it is important to consider the circumstances of each case, it is not the outcome of the case itself that is of prime importance, but the effect of admission or exclusion on our system of justice over time.

R. V. DUBOIS (1990)   **continued**

12. Through its decisions in specific cases, the Supreme Court has established that the administration of justice is not brought into disrepute by the admission of evidence where the detainee, advised of his rights, (a) provoked police haste by actively obstructing their investigation, or (b) failed to exercise his rights with reasonable diligence, or (c) on a breathalyzer charge, was unable to understand by reason only of impairment by alcohol. This last decision might govern as well any case where the detainee, told what his rights are, cannot understand or exercise them by reason only of the very impairment that is the gravamen of the offence charged. (pp. 95–197)

The term "without delay" has been read to mean as soon as reasonably possible, if it is not practicable to occur immediately. It is felt that police must allow the detainee access to a telephone, not just once, but as many times as are reasonably necessary to retain and instruct counsel.

Questioning must cease once a suspect has expressed a wish to consult counsel. An important case in this instance is *Manninen* (1987), 58 C.R. (3d) 97 (S.C.C.). Manninen had been arrested for armed robbery and had been read his rights based on a card provided to police officers once the Charter came into force. There had been a serious violation of his right to counsel, seen as being deliberate and flagrant under the circumstances. Here the court noted that the right to counsel is meant to afford the accused an opportunity to obtain advice as to how to exercise those rights. The court recognized there might be circumstances that demanded the police continue their investigation before it is feasible for an accused to contact a lawyer.

An accused can certainly waive his or her right to counsel. In the U.S. it is necessary for that waiver to be in writing; this is not the case in Canada. The waiver, however, must be clear and unequivocal, with indication of a full knowledge of the right waived and the effect of that waiver. The onus lies with the police to prove to the court that the accused waived a Charter right. In *Wills* (1992), 70 C.C.C. (3d) 529, 12 C.R. (4th) 58 (S.C.C.) the court ruled that if the prosecution is going to advance the position that an accused waived a constitutional right, it must establish on a balance of probabilities that:

1. There was consent, express or implied;
2. The giver of the consent had authority to give the consent in question;
3. The consent was voluntary and was not the product of police oppression, coercion, or other external conduct which negated the freedom to choose whether or not to allow the police to pursue the course of conduct requested;

4. The giver of the consent was aware of the police conduct to which he or she was being asked to consent;
5. The giver of the consent was aware of his or her right to refuse to permit the police to engage in the conduct requested; and
6. The giver of the consent was aware of the potential consequences of giving the consent.

A case where the police failed to prove that the accused waived the right to counsel was that of *Greig* (1987), 56 C.R. (3d) 229 (Ont. H.C.). The accused was re-interviewed by the police and not informed of his right to counsel; his earlier statement was read to him but his subsequent statement to the police was excluded.

In *Burlingham* (1995), 38 C.R. (4th) 265, 97 C.C.C. (3d) 385 (S.C.C.) it was determined that an accused cannot enter into a plea bargain without benefit of counsel. As Salhany has observed:

> Any offer of a plea bargain must be made to the accused's counsel or to an accused in the presence of his counsel, unless the accused has expressly waived the right to counsel. The officers engaged in the plea bargain process must act honourably and forthrightly. (p. 43)

The right to retain and instruct counsel must be exercised in private. The police must provide for this privacy as noted in *McKane* (1987), 58 C.R. (3d) 130 (Ont. C.A.).

PROTECTION AGAINST ARBITRARY DETENTION

Section 9 of the Charter guarantees that everyone will be accorded the right not to be detained or imprisoned arbitrarily. Such a detention would be capricious, frivolous, unwarranted, and contrary to normal standards. An officer must be able to give a reason for the detention of someone and that reason must be based on reasonable grounds. See *Storrey* (1990), 53 C.C.C. (3d) 316 (S.C.C.). The case of *Wilson* (1990), 56 C.C.C. (3d) 142 (S.C.C.) is appropriate to consider in some detail because it provides a useful context for the court's decision. The conviction was upheld in this case because the officer's actions were deemed appropriate under the circumstances. The police officer had stopped the appellant, Mr. Wilson, even though there was no reason to believe that he was doing anything wrong. The officer was conducting a floating spot check in an area near a small-town hotel, where the bars had just closed. He noted that the vehicle in question had out-of-province plates, and he did not recognize any of the three men in the front seat of the pickup truck. The court did not view this as being a random stop because while they might not be suitable grounds for stopping a vehicle in larger cities (like Toronto or Edmonton), they are acceptable in the context of a smaller, rural community.

The case of *R. v. Labine* (1987), 49 M.V.R. 24 (B.C. Co. Ct.) deals with the Charter issue of arbitrary detention or imprisonment, stemming from a police policy of arresting everyone suspected of impaired driving, rather than issuing those persons an appearance notice. (See box below.)

R. V. LABINE (1987), 49 M.V.R. 24 (B.C. CO. CT.)

Facts:

The appellant, Labine, was involved in a motor vehicle accident. Police arrested him for impaired driving and demanded that he accompany them for a breathalyzer test. There followed an arrest without a warrant and without Labine being given the benefit of his right to be summonsed or given an appearance notice. At this time, the Royal Canadian Mounted Police policy in Burnaby, British Columbia, was that all impaired driving suspects are arrested.

Decision:

Speaking on behalf of the court, Hogarth Co. Ct. J. (County Court Judge) made the following observations:

> As impaired driving is one of the offenses that comes within the preview [sic] of 450(2)(b), being punishable by summary conviction and indictment, the accused upon his arrest was entitled as of right to freedom from arrest without warrant unless some one or more of the contingencies mentioned in s. 450(2)(d) and (e) were prevalent. In my view, the police cannot deliberately adopt a policy to deprive the appellant of that right, notwithstanding that the officer might be acting lawfully and in the execution of his duties under s. 450(3). (p. 28)

This judgment indicates that broad police policies that may tend to impact substantially on individual rights and freedoms will not necessarily be seen as acceptable to the courts. Hogarth continues in his decision:

> The police, by a policy which transcends the circumstances of the appellant and applies to all persons who have the misfortune to be suspected of impaired driving in one of the largest R.C.M.P. detachments in Canada, have taken it upon themselves, without authority, to arbitrarily abrogate the right of the accused to be free of arrest without warrant in the absence of certain contingencies permitting them to do so, and to sustain a policy would be to invite police officers to disregard the right of the accused to be free of arbitrary detention and to do so with an assurance of impunity. (p. 31)

As a result of this ruling, Mr. Labine's appeal was allowed, and his earlier conviction was set aside.

However, in the case of *Ladouceur* (1990), 56 C.C.C. (3d) 22 (S.C.C.) the Supreme Court of Canada held that while provincial highway legislation authorizing the random stopping of vehicles to determine if the drivers have a valid driver's licence is random and arbitrary, it deals with an important issue, namely, the unlicensed driver.

In offering the court's judgment, Justice Cory provided the following sobering observations:

> To recognize the validity of the random check is to recognize reality. In rural areas it will be an impossibility to establish an effective organized programme. Yet the driving offences in these areas lead to consequences just as tragic as those that arise in the largest urban centres. Even the large municipal police force will, due to fiscal constraints and shortages of personnel, have difficulty establishing an organized programme that would constitute a real deterrent. (p. 42)

Furthermore, the existence of a routine check is something that can be countenanced by the political standards of our country:

> While the routine check is an arbitrary detention in violation of s. 9 of the Charter, the infringement is one that is reasonable and demonstrably justified in a free and democratic society. (p. 45)

In the case of *R. v. Simpson* (1993), 79 C.C.C. (3d) 482 (Ont. C.A.) the issue of arbitrary detention is again the subject of the court's scrutiny. (See box below.)

R. V. SIMPSON (1993), 79 C.C.C. (3D) 482 (ONT. C.A.)

Facts:

An officer stops a vehicle near a suspected "crack house" and asks the occupants to get out of the car. The officer notices a bulge in the pants pocket of one of the occupants and asks: "What's in your pocket?" The occupant replies: "Nothing." The officer proceeds to locate a baggie in the occupant's pants pocket containing a quantity of cocaine.

Decision:

In providing the judgment on behalf of the court, Doherty J.A. made the following observation, which should be seen as providing an important limit on the capacity of a police officer to stop a vehicle:

> This detention was a direct result of the stopping of a motor vehicle. The lawfulness of the detention depends on the police officer's authority to stop the vehicle. The officer's purpose in effecting the stop is, in turn, relevant to the lawfulness of that stop. Constable Wilkin candidly

R. V. SIMPSON (1993) continued

acknowledged that his decision to stop the motor vehicle had nothing to do with the enforcement of laws relating to the operation of motor vehicles. Nor did Constable Wilkin rely on any specific statutory authority . . . when he stopped the vehicle.

In my opinion, the "check-stop" cases decide only that stops made for the purposes of enforcing driving-related laws and promoting the safe use of motor vehicles are authorized by s. 216(1) of the Highway Traffic Act, even where those stops are random. These cases do not declare that all stops which assist the police in the performance of any of their duties are authorized by s. 216(1) of the Highway Traffic Act. (p. 489)

And at page 500:

Constable Wilkin interfered with the appellant's liberty in the hope that he would acquire grounds to arrest him. He was not performing any service-related police function and the detention was not aimed at protecting or assisting the detainee. It was an adversarial and confrontational process intended to bring the force of the criminal justice process into operation against the appellant. (p. 500)

Also, at page 507:

The seriousness of these constitutional violations is also clear. Constable Wilkin obviously considered that any and all individuals who attended at a residence that the police had any reason to believe might be the site of ongoing criminal activity were subject to detention and questioning by the police. This dangerous and erroneous perception of the reach of police powers must be emphatically rejected. Judicial acquiescence in such conduct by the reception of evidence obtained through that conduct would bring the administration of justice into disrepute. (p. 507)

CONSEQUENCES OF A BREACH OF THE CHARTER

The Charter provides a remedy to anyone whose rights have been denied. Section 24 of the Charter provides:

(1) Anyone whose rights and freedoms, as guaranteed by this Charter, have been infringed or denied may apply to a court of competent jurisdiction to obtain such remedy as the court considers appropriate and just in the circumstances.

(2) Where, in proceedings under subsection (1), a court concludes that evidence was obtained in a manner that infringed or denied any rights or freedoms guaranteed by this Charter, the evidence shall be excluded if it is established that, having regard to all the circumstances, the admission of it in the proceedings would bring the administration of justice into disrepute.

First of all, everyone has a right to make an application to a court for such a remedy. This must be a court of "competent jurisdiction." The court has a right to award such remedy as it considers appropriate and just under the circumstances. One remedy might be damages against a police officer. This occurred in the case of *Crossman* (1985), 12 C.C.C. (3d) 547 (Fed. T.D.) where the officer was assigned damages in the amount of $500 for denying the accused's right to counsel. Section 24 also provides a further remedy, i.e., the exclusion of evidence obtained as a result of a breach of a constitutional right, if it is proven that any admission of that evidence would bring the administration of justice into disrepute. If it is determined that the administration of justice has been brought into disrepute, the court *must* exclude the evidence.

A determination on whether evidence was gained in a manner that infringes upon the Charter can be found in the case of *Goldhart* (1996), 48 C.R. (4th) 297 (S.C.C.).

In the case of *Stillman*, [1997] 1 S.C.R. 607 the Supreme Court of Canada clarified its position on the idea of bringing the administration of justice into disrepute. Evidence would henceforth be classified as "conscriptive" or "non-conscriptive," depending on the manner in which it was obtained. When someone is compelled to incriminate herself or himself, such as when the body or bodily samples are required by the police, that evidence will be classified as conscriptive and amounts to "compelled" testimony. The Court determined that the security of the body was worthy of protection against state intrusion:

> In other words, even real evidence which would not have been obtained "but for" the participation of the accused, will affect the fairness of the trial and will be excluded. This trend has undoubtedly been based on the belief that it is fundamentally unfair for the state to obtain an evidentiary advantage by using the accused to incriminate himself. That unfairness, it is believed, would not be acceptable to right-thinking members of the community and would bring the administration of justice into disrepute. (p. 56)

An extremely significant case dealing with the Charter issue of arbitrary detention can be found in *Brown v. Durham Regional Police Force* (1999), 131 C.C.C. (3d) 1 (Ont. C.A.). In this instance, the police had established checkpoints targeting members of a motorcycle gang. Matters of highway safety were raised, along with concerns for the maintenance of public peace, investigation of other criminal activity, and intelligence gathering. It was determined that detentions were authorized under the *Highway Traffic Act*. (See box below.)

GENERAL POWERS OF ARREST WITHOUT A WARRANT

The powers of arrest of a private citizen and peace officer are defined in the *Criminal Code of Canada*, for all those offences created by the Parliament of Canada. As may be expected, a peace officer has all of the powers of arrest possessed by a private citizen, plus additional powers. The powers that exist with respect to provincial statutes are defined through legislation passed by individual provinces.

BROWN V. DURHAM REGIONAL POLICE FORCE (1999), 131 C.C.C. (3D) 1 (ONT. C.A.)

Facts:

Officers of the Durham Regional Police had directions that anyone driving a Harley-Davidson motorcycle or wearing the colours of the Paradise Riders or their insignia were to be stopped and required to produce licence, ownership and insurance documentation. Motorcycle gang members were also detained while information was checked through the Canadian Police Information Centre (CPIC). Biker gang members were videotaped and in some cases questioned.

Decision:

In expressing the court's decision, Doherty J.A. made the following observation:

> ... the police are entitled on a s. 216(1) stop to require drivers to produce their licences. That requirement is consistent with the highway safety concerns which underlie the power granted by the section. In addition to ensuring that the driver is properly licensed, the police may wish to identify the driver for other purposes. It may be, as in this case, that the police are interested in knowing the identity of all those who are connected with what they believe to be organized criminal activity. The gathering of police intelligence is well within ongoing police duty to investigate criminal activity ... I see no reason for declaring that a legitimate police interest beyond highway safety concerns should taint the lawfulness of the stop and detention. As the trial judge pointed out, known criminals should not be more immune from s. 216(1) stops than law abiding citizens who are not known to the police. (p. 15)

The essential requirement for police officers to conduct their public safety activities in an effective manner is recognized by the court, and motorcycle gang members cannot expect protections beyond those accorded ordinary citizens with respect to legitimate police activities.

Doherty J.A. proceeds to indicate that it is always better to have straightforward guidelines in place with respect to police powers. However, there is an expression of common sense that not all circumstances can be easily or readily predicted:

> Obviously, clear and readily discernible rules governing the extent to which the police can interfere with individual liberties are most desirable. The infinite variety of situations in which the police and individuals interact and the need to carefully balance important but competing interests in each of those situations make it difficult if not impossible, to provide preformulated bright-line rules which appropriately maintain the balance between police powers and individual liberties. (p. 24)

BROWN V. DURHAM REGIONAL POLICE FORCE (1999) **continued**

Of particular interest, in the context of proactive, problem-solving modes of modern policing, is Doherty's observation on the nature of law enforcement. The issue of crime prevention is essential for effective policing in Canada and the court explicitly acknowledges this point:

Not all law enforcement is reactive. The police duty to prevent crime and maintain public peace commands proactive measures on their part. Often those measures do not conflict with any individual rights and do not raise constitutional issues. Many facets of community based policing involve proactive measures taken with the full support and cooperation of the those affected by the measures.(p. 25)

And further:

Proactive policing is in many ways more efficient and effective than reactive policing. Where proactive steps do not collide with individual rights, that increased efficiency and effectiveness comes at no constitutional cost. Even where there is interference with individual rights, the societal gains are sometimes worth the interference. (pp. 25–26)

Section 494(1)(a) of the Code authorizes anyone to arrest without a warrant a person "whom he finds committing an indictable offence." From a legal perspective, "found committing" holds two meanings:

1. where the accused is surprised in the act of committing an offence; or
2. where the accused has been pursued immediately and continuously following the commission of an offence.

All offences under federal statutes may be prosecuted by indictment or on summary conviction. Provincial offences are only prosecuted by way of summary conviction. The *Criminal Code* defines whether an offence is an indictable one or punishable on summary conviction. However, some offences, such as impaired driving or the dangerous operation of a motor vehicle, may be punishable either on indictment or on summary conviction. These are known as hybrid or dual offences. In these instances, the Crown must decide how the offence will be prosecuted. In order to clarify matters for private citizens and peace officers, the *Interpretation Act*, section 34(1)(a) states that "if the enactment provides that the offender may be prosecuted for the offence by indictment" it will be deemed an indictable offence, until the Crown decides how it will proceed. This point is discussed in the case of *Huff* (1979), 50 C.C.C. (2d) 324 (Alta. C.A.). Therefore, a private citizen or a police officer could arrest without a warrant someone found committing a hybrid or dual offence (e.g., impaired driving) even if the Crown subsequently decides to proceed by way of summary conviction.

Section 494(1)(b) of the *Criminal Code* permits anyone to arrest without a warrant a person who is being pursued following the commission of a crime. A "criminal offence" is any indictable offence or summary offence under the *Criminal Code* or any federal statute. It does *not* include offences under a provincial statute. "Fresh pursuit" is defined in *Macooh* (1993), 22 C.R. (4th) 70, 82 C.C.C. (3d) 481 (S.C.C.):

> The courts at common law have always authorized pursuit into the home of the fugitive or a third party because they have considered that the flight of an offender is an act contrary to public order and should not be rewarded. Offenders fleeing from arrest should not be encouraged to seek refuge in their homes or those of third parties. (p. 61)

Section 495(1)(a) of the *Criminal Code* gives a peace officer the power to arrest a person who has committed an indictable offence. In this case, it is not necessary for the officer to actually see the accused or surprise the accused in the act of committing the offence. This section further gives a police officer the power to arrest someone without a warrant if the officer believes on reasonable grounds that the person has committed an indictable offence. The officer must be able to point to something factual to justify his or her belief that an indictable offence has been committed by the person being arrested. An example of where a peace officer's belief was not reasonable is found in *Gelfand v. C.P.R.* (1925), 44 C.C.C. 325 (Man. K.B.).

In the case of *R. v. Storrey* (1990), 53 C.C.C. (3d) 316 (S.C.C.) the issue of the legality of an arrest without a warrant was considered. (See box below.)

Police officer with a copy of the *Criminal Code of Canada*.

Credit: Mike Weaver, Media Relations, Kingston Police

R. V. STORREY (1990), 53 C.C.C. (3D) 316 (S.C.C.)

Facts:

In 1982, Americans driving near the Ambassador bridge that joins Windsor and Detroit were cut off and stopped by a vehicle. The driver and passenger of the vehicle attacked the Americans and all of them were slashed with a knife. The car was identified as possibly a 1973 or 1975 blue Thunderbird. Mr. Storrey was determined to be the owner of a 1973 blue T-bird. He was also found to have an extensive criminal record, including offences for violence.

Decision:

In this case, the court found that there were reasonable and probable grounds for an arrest, given these circumstances.

Section 495(1)(a) of the *Criminal Code* includes a power of arrest for a peace officer who believes, on reasonable grounds, that someone is about to commit an indictable offence. One illustration of this is the intoxicated person who is seen to be heading for her or his vehicle. The police officer may arrest this person without a warrant on the reasonable belief that she or he is planning to drive that car, thus committing an offence of impaired driving.

Another case dealing with an arrest without a warrant is *R. v. Macooh* (1993), 22 C.R. (4th) 70, 82 C.C.C. (3d) 481 (S.C.C.). (See box below.)

R. V. MACOOH (1993), 22 C.R. (4TH) 70, 82 C.C.C. (3D) 481 (S.C.C.)

Facts:

The accused was seen going through a stop sign. The police followed with emergency lights activated. The accused sped up and went through two more stop signs, then proceeded to the parking lot of an apartment building. The police followed and called for him to stop, but he ran into the building. The police called for him through the door of the

R. V. MACOOH (1993)   continued

apartment, but received no answer. They then entered the apartment and found him in bed. He was informed that he was under arrest for failure to stop for an officer. He refused to get dressed and when the officers were compelling him to dress, an altercation broke out. Their subsequent demand for a breath sample was refused. The accused was charged with impaired driving, failing to stop, failing to submit to a breathalyzer test, and assaulting a peace officer with intent to resist arrest.

Decision:

The trial judge found that entry to apprehend a person suspected of a breach of summary legislation in a provincial statute (as opposed to an indictable offence) was unlawful and therefore the arrest was also unlawful.

On behalf of the court, Lamer C.J.C. (Chief Justice of Canada) offers the following insight:

> . . . a right of entry to make an arrest in hot pursuit exists at common law, both for indictable offences and for other types of offence. In my opinion, there is no need to alter this rule. There are strong policy considerations against retaining the distinction between indictable offences and other categories of offence in determining the spatial limits on the power of arrest in hot pursuits . . . Most importantly, there is no logical connection between the fact that an offence falls in one or other of these categories and the need there may be to make an arrest in hot pursuit in residential premises. (p. 83)

One extremely important recent case dealing with the legality of an arrest without a warrant is *R. v. Feeney* (1997), 115 C.C.C. (3d) 129 (S.C.C.). In this case the police officer indicated that he did not believe that grounds for an arrest were present. It was furthermore determined that neither subjective nor objective grounds were present. As a result, the arrest was unlawful. (See box below.)

R. V. FEENEY (1997), 115 C.C.C. (3D) 129 (S.C.C.)

Facts:

Mr. Feeney, the accused, was charged with the murder of Frank Boyle, an 85-year-old man, after a local resident suggested to the police that Feeney had been seen in the vicinity of the deceased's home. The accused lived in a windowless trailer and had been out drinking the night before. The police officer knocked on the trailer door and announced that he was from the police. After getting no answer, the officer entered the trailer and woke up Mr. Feeney. Blood was seen splattered over the front of Feeney's clothing. He was read his rights, but no reference was made to the toll-free number that was available for this purpose. After Feeney indicated that he understood his rights, the officer questioned him about the blood and his clothing. The officer seized the accused's shirt and he was taken to the police station. Feeney was unsuccessful in contacting a lawyer and the police took a breathalyzer sample without informing him that he had a choice to accept or refuse this sample. Questioning of the accused continued for more than eight hours before he asked to speak to a lawyer. During questioning Feeney admitted to striking Boyle, and to stealing beer, cigarettes, and cash from the deceased. The cash was hidden under his mattress back at his trailer. The police obtained a search warrant authorizing the seizure of his shoes, cigarettes, and money.

During this period, Feeney had still not seen a lawyer, but questioning continued. Two days after his arrest, Feeney finally met with a lawyer.

Decision:

In presenting the court's decision Justice Sopinka, at page 132, makes the following observation that speaks to both the common law and the Charter:

> Even if the police met the common law standards, a warrantless arrest in the circumstances following a forcible entry is no longer lawful in light of the Canadian Charter of Rights and Freedoms. In general, the individual's privacy interest in the dwelling-house outweighs the interest of the police and warrantless arrests in dwelling-houses are prohibited.

When considering the sanctity of someone's home, police officers take on a special onus to respect the privacy of every individual. Justice Sopinka further indicates, at page 132:

> At common law a warrantless arrest following a forced entry into private premises is legal if: (a) the officer has reasonable grounds to believe that the person sought is within the premises; (b) proper announcement is made; (c) the officer believes reasonable grounds for the arrest exist; and (d) objectively speaking, reasonable and probable grounds for the arrest exist.

The judgment in this case provides a useful outline of the considerations that could, in fact, justify a warrantless search on private premises. In the case of Feeney, these conditions did not exist.

With regard to detention, it was deemed to have begun at the point where the police officer touched Mr. Feeney's leg and ordered him to get up from his bed. Feeney was not given a caution at this time and this was in violation of his section 10(b) rights under the Charter.

R. V. FEENEY (1997) **continued**

In looking at the matters that prompted the police to seek to question Mr. Feeney the court found that nothing served to raise reasonable and probable grounds that he was responsible for the murder of Boyle. Specifically, the court noted:

> Whether or not the appellant had been involved in two similar truck accidents, or might have stolen Boyle's truck, does not raise reasonable and probable grounds that he had murdered Boyle. This evidence may have pointed to the appellant as a suspect, but these facts without more do not justify an arrest. When the police entered the trailer, objectively reasonable and probable grounds for an <u>arrest</u>, as opposed to grounds for prima facie suspicion, did not exist. (p. 151)

At page 156:

> In cases of hot pursuit, society's interest in effective law enforcement takes precedence over the privacy interest and the police may enter a dwelling to make an arrest without a warrant. However, the additional burden on the police to obtain a warrant before forcibly entering a private dwelling to arrest, while not justified in a case of hot pursuit, is, in general, well worth the additional protection to the privacy interest in dwelling-houses that it brings.

At page 159:

> The circumstances surrounding the police entry into the trailer were similar to those following any serious crime: a dangerous person is on the loose and there is a risk that he or she will attempt to destroy evidence linking him or her to the crime. To define these as exigent circumstances is to invite such a characterization of every period after a serious crime.

The appeal by Feeney was allowed, his conviction was set aside, and a new trial was ordered.

A dissenting view was presented by Justice L'Heureux-Dubé, that speaks to the challenge of policing in Canada, and to some of the frustration occasionally experienced by police officers:

> The police were in the process of investigating a serious crime, one which had recently been committed and involved a savage, physical beating inflicted on a helpless victim for no apparent reason. Given the brutality of the murder scene and the seeming randomness of the act, there is little doubt that the police felt obliged to act quickly in order to prevent any further violence of the nature in the community. For this foresight, they should be commended, not rebuked. (pp. 204–205)

Section 31 of the *Criminal Code* allows a peace officer to arrest anyone found committing a breach of the peace, as long as the officer witnesses this breach. The officer may also arrest anyone whom she or he believes, on reasonable grounds, is about to join in or renew this breach of the peace. See *Hayes v. Thompson* (1985), 44 C.R. (3d) 316 (B.C.C.A.). There is a natural extension relevant to the peace officer's responsibility to prevent offences.

Section 495(1)(c) authorizes a peace officer to arrest without a warrant a person when there are reasonable grounds to believe that a warrant of arrest or committal for that person is in force within the jurisdiction in which the person is found.

SPECIFIC POWERS OF ARREST

The *Criminal Code* contains specific arrest powers without a warrant for a peace officer or private citizen as follows:

- An owner of property or someone lawfully authorized to be in possession of it is entitled to arrest a person whom he or she finds committing a criminal offence on or in relation to that property (section 494(2)). For example, one person who finds another person trying to steal a car from his or her garage, or trying to remove a stereo from the car, may make an arrest.
- A peace officer may arrest without a warrant someone found keeping a common gaming house and any person found on those premises (section 199(2)).
- A peace officer may also arrest without a warrant an accused who (a) is believed, on reasonable grounds, to be contravening or is about to contravene any summons, appearance notice, promise to appear, undertaking, or recognizance issued to her or him, or (b) has committed an indictable offence, after one has been issued to her or him (section 524(2)).

LIMITATIONS ON POWERS OF ARREST

Section 495(2) of the *Criminal Code* places a duty *not* to arrest without a warrant under the following circumstances:

- Where an indictable offence is within the absolute jurisdiction of a provincial court judge;
- Where an offence is a dual or hybrid offence;
- Where an offence is punishable on summary conviction; or
- Where it is likely that the accused will appear for trial, and therefore satisfy the public interest.

Section 495(3) of the *Criminal Code* explains that:

Notwithstanding subsection (2), a peace officer acting under subsection (1) is deemed to be acting lawfully and in the execution of his duty for the purposes of

(a) any proceedings under this or any other Act of Parliament, and
(b) any other proceedings, unless in any such proceedings it is alleged and established by the person making the allegation that the peace officer did not comply with the requirements of subsection (2).

This section is explained further in *Adams* (1972), 21 C.R.N.S. 257 (Sask. C.A.). The case of *McKibbon* (1973), 12 C.C.C. (2d) 66 (B.C.C.A.) is also valuable in this context. The onus is on the person who claims that the officer did not comply with section 495(2). Only when this is proved does the arrest become unlawful and can the officer be rendered liable for damages in a civil suit.

PROCEDURE TO BE FOLLOWED WHERE THE PEACE OFFICER DOES NOT ARREST

If a peace officer elects not to arrest someone, it remains possible for that officer to prepare and serve the accused with something called "an appearance notice" (see the *Criminal Code*, section 496) or, the officer may appear before a justice of the peace and lay an information charging the accused with an offence (section 504). In the latter case, the officer can apply to the justice for a summons or a warrant that will compel the accused to appear in court at a specified time and place to answer to the charge (section 507). Once the information is sworn, the officer cannot issue a new appearance notice but may only apply for a summons or warrant, as per *Curlew* (1981), 64 C.C.C. (2d) 211 (Nfld. C.A.). An appearance notice must contain the following information, according to section 501(1) of the *Criminal Code*:

1. Name of the accused;
2. Substance of the charge;
3. Time when and the place where the accused is to attend court to answer the charge; and
4. Text of sections 145(5) and (6) and section 502 of the Code.

The notice may also require the accused to appear at a stated time and place (under section 501(3)) for the purposes of the *Identification of Criminals Act*, R.S.C. 1985, c. I-1.

An appearance notice should be signed in duplicate by the accused and one of the copies retained by the accused. If the accused refuses to sign the appearance notice, it may remain valid if confirmed by the justice of the peace.

DUTIES AFTER ARREST

The *Criminal Code of Canada* imposes certain duties on a person who does arrest someone. A private citizen is obligated to deliver the arrested person "forthwith" to a peace officer as per section 494(3). Also, an officer who receives an arrested person, or who makes an arrest, has the following duties to perform:

The officer may release the accused unconditionally (sections 503(1)(d) and 3(a)) or conditionally (section 503(2)), where the officer is satisfied that the accused should be released. If the person is released conditionally, the officer must issue the accused an appearance notice (if a peace officer) or require the accused to give a promise to appear or enter into a recognizance (if an officer in charge).

The officer must release the accused unconditionally if satisfied that the accused's continued detention is no longer necessary to prevent the commission of an indictable offence. This arises where the officer has arrested a person whom she or he believed on reasonable grounds was about to commit an indictable offence.

A peace officer must release an accused who has been charged with an indictable offence within the absolute jurisdiction of a provincial court judge, a hybrid or dual offence, or summary offence unless he believes that it is necessary to detain the accused in the public interest or to ensure his appearance.

If the offence is more serious, or the accused does not qualify to be released, the peace officer may deliver the accused to an officer in charge of a lock-up or to a justice of the peace. If the accused is delivered to an officer in charge, that person is required to release the accused for an indictable offence punishable by imprisonment of five years or less, unless the officer is satisfied that it is necessary to detain the accused in the public interest or to compel the accused's appearance (*Criminal Code*, section 498(1)). The officer in charge must release the accused as soon as practicable if a decision to release is made.

If the peace officer decides that the accused should be detained, she or he must take the accused before a justice of the peace without unreasonable delay, and not beyond 24 hours following the accused's arrest if a justice is available. If a justice is not available within 24 hours, the officer must take the accused before a justice as soon as possible (*Criminal Code*, sections 503 (1)(a) and (b)).

A peace officer has no power to release an accused who has committed an offence that is within the absolute jurisdiction of a judge of the superior court such as murder, piracy, or treason or an attempt or conspiracy to commit one of those offences. In such instance, the accused must be taken before a justice of the peace. Only a judge of the superior court has jurisdiction to release the accused.

A peace officer, or an officer in charge, has no power to release a person who has been arrested without a warrant for an indictable offence committed in Canada but outside the province where the accused was arrested (*Criminal Code*, section 503(3), amended 1997, c. 18, s. 55(3), proclaimed May 14, 1997). In such instance, the person must be taken before a justice of the peace in the jurisdiction where the arrest was made, within the time specified in section 4 (above). If the justice is not satisfied that the person arrested is the one alleged to have committed the offence, she or he must release that person. If the justice is so satisfied, then she or he may, if the Crown consents, release the accused unconditionally, or upon an undertaking or recognizance with or without terms. If the Crown does not consent, the justice must remand the accused into custody to await a warrant for his or her arrest from the province where the accused is wanted. If the warrant is not executed within six days, the justice must release the accused.

LAYING OF AN INFORMATION

Where an accused has been released by a peace officer or an officer in charge, pursuant to an appearance notice, promise to appear, or recognizance, an information must be laid before a justice of the peace as soon as possible, and at least before the date set for the appearance of the accused (*Criminal Code*, section 505). An information is a written complaint upon oath by someone stating that she or he has personal knowledge or reason to believe that the person named in the information has committed an offence. The information is generally laid by a peace officer, although there is no requirement that this be so.

In summary conviction matters, the court is entitled to award costs against the unsuccessful party, whether it is the prosecutor or the defendant. Thus, the general practice is for the police to let the complainant lay the information personally, thereby avoiding the risk of having to pay costs.

The justice of the peace who receives an information laid before him is required to hear and consider the following, *ex parte* (i.e., in the absence of the defendant):

- Allegations of the informant; and
- Evidence of witnesses, where he or she considers it desirable or necessary to do so (*Criminal Code*, section 508(1)(a)).

If the justice considers that a case has been made out, he or she may confirm the appearance notice, promise to appear, or recognizance and endorse the information. The justice may also, where he or she considers that a case for doing so is made out, cancel the appearance notice, etc., and issue a summons or warrant (*Criminal Code*, section 508(1)(b)). In other words, the justice can now override the decision of the officer to release the accused and force a show-cause hearing.

ARREST WITH A WARRANT

While the *Bail Reform Act* re-affirmed the power of a justice of the peace to issue a warrant for the arrest of someone suspected of having committed a criminal offence, a restriction has been imposed: a justice must not issue a warrant for the arrest of someone whose appearance for trial can be compelled by a summons, unless the justice has reasonable grounds to believe that it is necessary in the public interest to do so (*Criminal Code*, sections 507(4) and 512(1)). However, before a warrant can be issued, an information must be laid before a justice alleging that a criminal offence has been committed.

The contents of the warrant and the manner of executing the warrant are set out in section 511 of the *Criminal Code*. It must name or describe the accused, state briefly the charge against her or him, and be signed by the justice. The warrant will be directed to the peace officers generally within the jurisdiction of the court issuing it and will order the accused to be arrested and brought before the justice of the peace or some other justice having jurisdiction to hear the charge.

Any peace officer to whom the warrant is directed may arrest the accused either in the territorial jurisdiction of the justice issuing the warrant, or anywhere in Canada if the officer is freshly pursuing him or her (*Criminal Code*, section 514). As soon as the accused is validly arrested the warrant is seen to be executed.

Section 20 of the *Criminal Code* authorizes a justice to issue a warrant or summons on a holiday; it may also be executed by a peace officer on one of those days.

WHEN MAY A WARRANT BE ISSUED?

A warrant may also be issued under the following circumstances:

* Where an accused fails to appear in court in answer to a summons (*Criminal Code*, section 512(2)(a));
* Where an accused fails to attend court (a) in accordance with a confirmed appearance notice, promise to appear, or recognizance entered into before an officer in charge (*Criminal Code*, section 512(2)) or (b) for the purpose of identification under the *Identification of Criminals Act* (*Criminal Code*, section 502);
* Where the accused is evading service of a summons (*Criminal Code*, section 512(2)(c));
* Where there are grounds to believe that the accused has violated or is about to violate a promise to appear, etc. (*Criminal Code*, section 524(1)(a); and
* Where the accused has committed an indictable offence after her or his release (*Criminal Code*, section 524(1)(b)).

CARRYING THE WARRANT

According to the common law and section 29(1) of the *Criminal Code*, a peace officer who executes a warrant has a duty to have the warrant with him or her, where it is feasible to do so, and to produce it for the inspection of the accused, if asked.

However, this provision is really an anachronism in view of the general size of Canadian cities and towns, and it may not be possible for an officer to produce the warrant immediately. There is a need to take into account the difficulties imposed on officers because of the distances they might be required to travel, and the time constraints such travel might impose.

Section 29(2) also makes it

> . . .the duty of everyone who arrests a person whether with or without a warrant, to give notice to that person, where it is feasible to do so, of
>
> (a)   the process or warrant under which he makes the arrest; or
> (b)   the reason for the arrest.

If the officer is required to give the accused notice of both the process or warrant under which he makes the arrest and the reasons for the arrest, is it sufficient if she or he only does one of those things? Furthermore, what is meant by "notice . . . of the process or warrant"? In the case of *Gamracy* (1973), 12 C.C.C. (2d) 209 (S.C.C.) it was held by the court that the officer did not have to do both. It was seen as being sufficient if the accused was given either notice of the process or warrant or of the reason for the arrest in order to comply with section 29(2). If the officer told the accused that the reason for the arrest was the existence of the outstanding warrant it is not necessary to produce the warrant or to ascertain its whereabouts:

> The Gamracy case is probably no longer good law in Canada since the enactment of section 10(a) of the Canadian Charter of Rights and Freedoms. That section guarantees everyone the right "on arrest or detention to be informed promptly of the reasons thereof". In other words, the Charter requires a police officer to inform anyone whom he arrests of the reasons for that arrest, whether or not he carries the warrant. (p. 76)

EXECUTION OF THE WARRANT WHERE THE ACCUSED IS IN ANOTHER PROVINCE

If a warrant cannot be executed because the accused is in another province, then any officer who wishes to execute the warrant in that province must apply to a justice of the peace who has jurisdiction in that province to endorse the warrant (*Criminal Code*, section 528). The officer is required to prove the issuing justice's signature either upon oath or by affidavit. A recent amendment to the *Criminal Code* (section 528(1.1)) allows the officer to produce, instead of the original, a copy of an affidavit or warrant submitted by a communication which produces a writing, such as a facsimile. If the justice is satisfied with the proof of signature he or she will then endorse the warrant in accordance with Form 28 of the *Criminal Code*. This endorsement will then be sufficient authority to the peace officer(s) named in the warrant or in the jurisdiction where it is endorsed to arrest and take the accused back to face the charge.

SUMMONS

A summons is an order of a justice of the peace addressed to an accused directing her or him to appear at a specific time and place to answer the charge as set out in the summons. As in the case of a warrant, an information must be laid before a justice of the peace can issue a summons (*Criminal Code*, section 507). Service of the summons must be made upon the accused personally. If the accused cannot be found conveniently, service can be made by leaving it at her or his last or usual place of abode with someone who appears to be at least 16 years of age (*Criminal Code*, sections 509(2) and (3)).

QUESTIONING A SUSPECT

Common law has recognized the right of a police officer to question a person without arresting that person (see *Bazinet* (1986), 25 C.C.C. (3d) 273 (Ont. C.A.)). The common law has also recognized that the suspect is entitled to ignore the officer's questions and to walk away. This places the officer in a position where he or she has a choice: to let the suspect walk away or to arrest the suspect. An arrest can occur only if the officer has reasonable grounds to believe that the person has committed an indictable offence.

In discussing the care required by police officers when asking someone to accompany them to a police station for questioning, Salhany (1997) makes reference to the case of *Forsyth v. Goden* (1895), 32 C.L.J. 288. In that case, it was determined that if someone, known to be a police officer, takes charge of someone, and that person reasonably thinks that he or she is under arrest by means of the officer's conduct, then there is an arrest.

ARREST OF WRONG PERSON OR ON FAULTY GROUNDS

A police officer who acts in good faith and upon reasonable grounds is protected from criminal responsibility by section 28 of the *Criminal Code*:

28.(1) Where a person who is authorized to execute a warrant to arrest believes, in good faith and on reasonable grounds, that the person who he arrests is the person named in the warrant, he is protected from criminal responsibility in respect thereof to the same extent as if that person were the person named in the warrant.

(2) Where a person is authorized to execute a warrant to arrest,
   (a) every one who, being called on to assist him, believes that the person in whose arrest he is called upon to assist is the person named in the warrant, and
   (b) every keeper of a prison who is required to receive and detain a person who he believes has been arrested under the warrant,

is protected from criminal responsibility in respect thereof to the same extent as if that person were the person named in the warrant.

A similar protection exists in section 25 of the *Criminal Code*. Thus an officer who arrests a person with or without a warrant, and discovers later that her or his grounds are faulty, will be protected from criminal responsibility so long as she or he acted on reasonable grounds.

In *Greffe* (1990), 75 C.R. (3d) 257 (S.C.C.), the Supreme Court of Canada excluded evidence of heroin that was discovered during a rectal search, largely because the accused had been arrested for outstanding traffic warrants, clearly an unconnected matter.

SEARCHING AN ARRESTED PERSON

Canada does not support the "stop and frisk" approach to law enforcement, which has been the subject of much controversy in the United States. Neither is there a right under Canadian law or common law to automatically search someone who has been detained for questioning. The right to search someone is only subsequent to an arrest. The purpose of the search is to locate further evidence relating to the charge upon which the person has been arrested or to locate any item(s) that might assist the accused to escape from custody or permit him to cause any violence. For a clarification of this aspect see the case of *Gottschalk v. Hutton* (1921), 36 C.C.C. 298 (Alta. C.A.).

The cases of *Morrison* (1987), 58 C.R. (3d) 63 (Ont. C.A.) and *Miller* (1987), 38 C.C.C. (3d) 252 (Ont. C.A.) established that a police officer does not need to have reasonable grounds to conduct a search for weapons following an arrest. But the authority to search is not unlimited as discussed in *Cloutier v. Langlois* (1990), 74 C.R. (3d) 316 (S.C.C.) where it was determined that the power to search is governed by three considerations:

1. The power to search does not impose a duty. Police officers have some degree of discretion in conducting a search. Occasions may arise when the police do not see fit to conduct a search, while satisfying the law, and being conscious of public and officer safety.
2. A search must be conducted for a valid objective in pursuit of the ends of criminal justice. The conduct of the search must not be unrelated to the objectives of the proper administration of justice. This would be the case if the search was conducted to intimidate, ridicule, or pressure the accused in order to obtain an admission.
3. A search must not be conducted in an abusive manner. The use of physical and/or psychological constraint should be proportionate to the objectives sought and the other circumstances of the case.

An instance where a complete search might not be warranted is in the case of an impaired driver. However, if the police are concerned about safety in this instance they have every right to conduct such a search, or, if the suspect is actively attempting to hide a bottle of liquor, as in the case of *Lerke* (1986), 24 C.C.C. (3d) 129 (Alta. C.A.), they likewise have a right to conduct a search.

The case of *Simmons* (1988), 66 C.R. (3d) 297 (S.C.C.) outlines three types of searches:

1. A pat-down search or frisk of outer clothing;
2. A body search, which involves the removal of articles of clothing in order to examine the person's body, or parts of the body; and
3. A body cavity search or probe, which may involve x-rays, emetics, or other highly intrusive measures or techniques requiring the assistance of medical doctors or specially trained personnel.

Simmons emphasizes the point that the greater the degree of intrusion on the person, the higher the degree of justification for the intrusion, and the greater the degree of constitutional protection accorded that person.

While the police are also permitted to conduct a search of the immediate surroundings when searching a person, this does not mean that the officer can conduct an exhaustive search of an entire building, adjacent buildings, vehicles, etc., to locate further evidence against the accused.

Section 101(1) of the *Criminal Code* does authorize a police officer to search a person without a warrant where the officer believes on reasonable grounds that an offence has been or is being committed with respect to restricted or prohibited weapons, firearms, or ammunition. The officer is then entitled to seize anything by means of or in relation to which he believes the offence has been committed.

Section 11(5) of the *Controlled Drugs and Substances Act*, S.C. 1996, c. 19, proclaimed May 14, 1997, permits a peace officer to enter a building with a search warrant and to search any person found in that place if she or he has reasonable grounds to believe that the person is in possession of any controlled substance, precursor, property, or thing set out in the warrant.

ARRESTING A YOUNG OFFENDER

The *Young Offenders Act*, R.S.C. 1985, c. Y-1, replaced the *Juvenile Delinquents Act* in 1985, and applies to any young person who is between the ages of 12 and 17 years. It provides for particular treatment for such persons as offenders. It has been the subject of much controversy over the last several years and is currently in the process of being amended to reflect an evolving understanding of the rights and responsibilities of young offenders in the context of criminal activity. The arrest of a young person involves some important considerations, including the nature of the offence and the age, size, and conduct of the young person.

A young offender who is arrested with or without a warrant may be taken before a justice of the peace to be dealt with prior to a plea. In all other matters, however, such as bail, the receiving of a plea or trial may only take place before a Youth Court judge.

Police officers must ensure that young offenders are not detained in a facility where adults are being held, with the following exceptions: A Youth Court judge may order a young person detained with adults if the judge concludes that it is not safe for the offender or others to detain him or her in a place of detention for young persons, or if there is no available place for detention of young persons within a reasonable distance.

TRESPASS IN MAKING AN ARREST

The right to search someone's house is considered an extraordinary remedy that can only be applied when there is a clear provision in law that allows it. This was established in an early case known as *Semayne's Case* (1604), 77 E.R. 194. An invasion of privacy is considered a trespass and strictly speaking no one has the right to enter private property without the owner's consent, or without strict adherence to some lawful authorization.

Therefore, a police officer has no right to enter private premises unless she or he holds a valid warrant, or is acting in accordance with some statutory provision. If the officer does not have a warrant or is not authorized by a specific statutory provision, the owner may view the officer as being a trespasser and may use whatever force necessary to remove the officer, as was the case in *Eccles v. Bourque* (1974), 19 C.C.C. (2d) 129 (S.C.C.). In this regard, a police officer holds no greater authority than another member of the public.

In the case of *Landry* (1986), 50 C.R. (3d) 55 (S.C.C.) the court indicated that a police officer could enter a private home to arrest someone if the officer could answer "Yes" to the following three questions:

1. Is the offence in question indictable?
2. Has the person who is the subject of the arrest committed the offence in question, or does the peace officer, on reasonable and probable grounds, believe that the person has committed or is about to commit the offence in question?
3. Are there reasonable and probable grounds to believe that the person sought is within the premises?

Should all of these questions receive an affirmative response, the officer would then be required to provide the owner (or occupant) with a proper announcement that *must* include the following elements:

1. Notice of authority as a peace officer;
2. Notice of the purpose; and
3. Request to enter the premises.

On this last point, the request to enter must be denied before the officer can proceed, unless there are unusual circumstances, such as the necessity of saving someone from death or injury, or to prevent the destruction of evidence or if the officer is in hot pursuit.

On May 22, 1997, however, the Supreme Court of Canada announced that the *Landry* test for warrantless searches no longer applied, except in cases of "hot pursuit." The *Feeney* case (1997), 115 C.C.C. (3d) 129 (S.C.C.) (discussed above) outlined the court's view that the privacy interest of the homeowner now outweighed the interest of the police and that warrantless searches for persons in a home are no longer permissible. The court was of the view that the privacy rights of the Charter demanded that the police, in general, should obtain prior judicial authorization before any entry into a private dwelling-house to arrest a person. Furthermore, a warrant for entering a private dwelling-house to arrest someone will only be authorized if there are reasonable grounds to believe that the person will be found on those premises.

There is currently no provision in the *Criminal Code* authorizing a judge to issue an arrest warrant allowing a police officer to enter a person's home to carry out an arrest. The Supreme Court did recognize this point and also that the absence of such a provision could have an impact on the common law power of arrest. In any event, the court felt its absence should not be allowed to defeat a constitutional right of the individual. An important issue that the court did not address in this case involves matrimonial disputes. Calls to the police for assistance frequently involve conflicts between couples (married or common-law). These situations can typically be quite violent and emotional levels are high. *Feeney* left the question of allowing for exceptional "emergency" circumstances until another time. It would appear, however, that an officer would be entitled to enter the premises of a homeowner without a warrant where an emergency existed and the officer believed, on reasonable grounds, that it was necessary to do so in order to preserve life or prevent serious injury to one of the parties. For a discussion of this point it is useful to consider the case of *Custer* (1984), 12 C.C.C. (3d) 372 (Sask. C.A.).

Police officers are placed in a difficult situation under these circumstances. Salhany has remarked:

> Faced with this situation, how can the officer determine whether he is entitled to remain, particularly where he does not have an opportunity of finding out who is the legal owner? This issue becomes even more difficult when one examines various provincial laws dealing with the property rights of legal and common-law spouses. (p. 93)

It has been suggested that in such circumstances, the police officer invite the party who has made the complaint to leave the premises with the officer. If there is danger, however, the police officer may stay in the premises if she or he has the permission of one occupier of the house, as was decided in *Sharpe* (1983), 9 W.C.B. 341 (N.W.T.S.C.).

ASSISTING A PEACE OFFICER

A police officer may call on anyone to assist him or her in carrying out an arrest, according to the common law. This common law duty is recognized by the *Criminal Code*, under section 129(b), making it an indictable offence to:

> . . . omit, without reasonable excuse, to assist a public officer or peace officer in the execution of his duty in arresting a person or in preserving the peace, after having reasonable notice that he is required to do so . . .

REFUSAL OF CITIZEN TO IDENTIFY SELF

A police officer must not use force to compel someone to identify himself or herself. If the officer does use force, that officer is guilty of a criminal assault and may be liable to civil damages. This matter is addressed in the case of *Koechlin v. Waugh and Hamilton* (1957), 118 C.C.C. 24 (Ont. C.A.) where a private citizen was arrested when he refused to disclose his identity to the police. (See box below.)

KOECHLIN V. WAUGH AND HAMILTON (1957), 118 C.C.C. 24 (ONT. C.A.)

Facts:

Two youths about 20 years old, in Scarborough, Ontario, were coming home from a movie about midnight. Two plainclothes officers travelling in a police cruiser stopped these young men and asked for their identification. One of the young men complied, but the other one, Koechlin, refused until the officers had identified themselves. One officer produced a badge but the plaintiff continued to refuse to identify himself and a scuffle ensued. He was placed in the police car and not given any reason for his arrest. Koechlin was taken to the police station and told he would be charged with assaulting a police officer. The sergeant did not provide details of the incident to the young man's father and he was refused permission to see his son. The charge was later heard and dismissed.

KOECHLIN V, WAUGH AND HAMILTON (1957)   continued

Decision:

In providing the court's decision, Laidlaw J.A. noted the following:

A police officer has not in law an unlimited power to arrest a law-abiding citizen. The power given expressly to him by the Criminal Code to arrest without warrant is contained in s. 435, but we direct careful attention of the public to the fact that the law empowers a police officer in many cases and under certain circumstances to require a person to account for his presence and to identify himself and to furnish other information, and any person who wrongfully fails to comply with such lawful requirements does so at the risk of arrest and imprisonment. None of these circumstances exist in this case...

We do not criticize the police officers in any way for asking the infant plaintiff and his companion to identify themselves, but we are satisfied that when the infant plaintiff, who was entirely innocent of any wrong-doing, refused to do so, the police officer has no right to use force to compel him to identify himself. (pp. 26–27)

Section 129(a) of the *Criminal Code* makes it an offence to wilfully obstruct police officers in the execution of their duties. The question then may be asked: is it an obstruction to refuse to identify oneself to a police officer who is carrying out his or her general law enforcement duties? Lord Parker, Chief Justice of England, in *Rice v. Connolly* (1966), 2 Q.B. 414 (C.A.) at 419, felt it was not and expressed himself accordingly, as follows:

It seems to me quite clear that though every citizen has a moral duty or, if you like, a social duty to assist the police, there is no legal duty to that effect, and indeed the whole basis of the common law is the right of the individual to refuse to answer questions put to him by persons in authority, and to refuse to accompany those in authority to any particular place; short, of course, of arrest.

An interesting case in this area is *Moore* (1979), 13 C.C.C. (2d) 83 (S.C.C.), where an accused, riding a bicycle, went through a red light. A police officer asked the person to stop, but that person did not comply. When the accused eventually did stop he refused to identify himself and the officer charged him with wilful obstruction. The trial judge directed a verdict of acquittal on the basis that the accused was not subject to the provisions in the British Columbia *Motor Vehicle Act* that made it an offence for the operator of a "motor vehicle" to refuse to stop when signalled by the police and to identify herself or himself when requested. The majority of the Supreme Court of Canada held that the trial judge erred in this regard. It ruled that the accused was clearly not in breach of the section of the statute referred to, since a bicycle is not a "motor vehicle." But under section 17 of the British Columbia *Police Act, 1974*, there is a duty on every municipality to set up a police force that will enforce municipal by-laws, the criminal law, and the laws of the province. Section 30 of the Act states that a police officer has jurisdiction to carry out various "powers, duties, privileges, and responsibilities . . . at law or under any Act." Therefore, a police officer is permitted to arrest an accused whom he or she finds committing a summary offence to "establish the identity of the person." The accused in this case was, then, obstructing the officer in the performance of his duty.

In *Guthrie* (1982), 28 C.R. (3d) 395 (Alta. C.A.), the court stated that the general rule is that a person is not required to identify himself or herself. There is no right for the officer to ask a person to identify himself or herself unless the officer actually observed the person committing an offence known to law.

CONSEQUENCES OF AN ILLEGAL ARREST

A defence, on the basis of unlawful or illegal arrest, can be presented to a charge of escaping lawful custody, assaulting a peace officer in the execution of his or her duty, or obstructing or resisting a peace officer in the execution of his or her duty. It must be established that a police officer was acting under a duty (see *Johansen*, [1947] 4 D.L.R. 337 (S.C.C.)). If a police officer makes an unlawful arrest, it follows that there is a common law right to resist such an arrest. This point is discussed in the case of *Christie v. Leachinsky* (1947), A.C. 573 (H.L.).

ARREST UNDER PROVINCIAL STATUTES AND BY-LAWS

Only Parliament can pass laws relating to the criminal law. This is established by section 91(27) of the *Constitution Act* of 1867. The provinces have the right to create and enforce legislation within their jurisdiction (e.g., *Highway Traffic Act*, *Liquor Licence Act* in Ontario). These offences may only be prosecuted by way of summary conviction, not by indictment.

In Ontario, the *Provincial Offences Act*, R.S.O. 1990, c. P-33, governs the trial of provincial offences. Sections 144(2) and 145 authorize arrests, without a warrant, under the following circumstances:

144(2). A police officer may arrest without warrant a person for whose arrest he or she has reasonable and probable grounds to believe that a warrant is in force in Ontario.

145. Any person may arrest without warrant a person who he or she has reasonable and probable grounds to believe has committed an offence and is escaping from and freshly pursued by a police officer who has lawful authority to arrest that person, and where the person who makes the arrest is not a police officer, shall forthwith deliver the person arrested to a police officer.

The following provides a listing of some statutes that include powers of arrest and briefly outlines the rationale for each power:

ONTARIO *HIGHWAY TRAFFIC ACT*

Section 217(2) and (3) relating to s.9(1) making a false statement in an application . . .

s. 12(1)(a) defacing or altering a number plate or evidence of validation or permit;

s. 12(1)(b) using or permitting the use of a defaced or altered number plate . . .

s. 12(1)(c) removing a number plate without authority;

s. 12(1)(d) using or permitting the use of an unauthorized number plate;

s. 12(1)(e) using or permitting the use of validation upon a number plate not authorized by the Ministry;

s. 12(1)(f) using or permitting the use of a number plate or evidence of validation not authorized by the Ministry;

s. 13(1) exposing wrong number on plate;

s. 33(3) unable or refuses to identify oneself on failure to produce a licence;

s. 47(5) unlawful application for, procuring of or possession of a motor vehicle permit plate while suspended or cancelled;

s. 47(6) unlawful application for, procuring of or possession of a driver's licence while suspended or cancelled;

s. 47(7) applying or procuring a CVOR certificate while suspended;

s. 47(8) operating a commercial motor vehicle while restricted, or without a permit or certificate, or while suspended;

s. 51 operating a motor vehicle where a permit is suspended or cancelled;

s. 53 driving while licence is suspended;

s. 130 careless driving;

s. 172 racing on a highway;

s. 184 removing or defacing notices or obstructions lawfully placed on a highway;

s. 185(3) refusal by pedestrian on highway in contravention of regulation or by-law to obey request of police officer to accompany him to nearest intersecting highway not prohibited for use;

s. 200(1)(a) failure to remain at the scene of an accident; [and]

s. 216(1) refusal of motorist to stop vehicle when required by police officer in lawful execution of duties and responsibilities.

ONTARIO *LIQUOR LICENCE ACT*

Section 48 of this statute authorizes a police officer to arrest without a warrant any person whom the officer finds apparently in contravention of the Acts or regulations and who refuses to provide her or his name and address, or where there are reasonable grounds to believe that the name and address provided are false.

ONTARIO *GAME AND FISH ACT*

Section 10 of this statute authorizes a police officer to arrest without process any person found committing a contravention of the Act or its regulations. Section 11 authorizes the officer to enter upon or pass through or over private lands to discharge these duties.

The important point is that the officer has authority only when he or she is in a position to say that he or she "finds [someone] committing" a specific offence. Summary offences are clearly not as serious as indictable offences; therefore, officers must bear this difference in mind when effecting arrests in these circumstances.

MUNICIPAL BY-LAWS

The question is whether or not the Ontario *Municipal Act* provides the municipality with the authority to pass by-laws that include arrest provisions. Ontario, for example, has no such power extended through its *Municipal Act*.

In the case of *R. v. Polashek* (1999), 134 C.C.C. (3d) 187 (Ont. C.A.), the accused was charged with possession of narcotics and possession for the purpose of trafficking, after being lawfully stopped for a *Highway Traffic Act* violation. The officer noted the "strong odour of marijuana" but saw no smoke. The accused denied the presence of dope. The officer noted that the smell, the accused's denial, the area, and the time of night provided the officer with grounds for an arrest. Upon arrest, the officer continued the search and found more than $4,000 in the accused's pocket and in shoe boxes with wrapped bags of weed, a scale, rolling tobacco, etc.

Rosenberg J.A. (Justice of Appeal) has ruled:

> In my view, the search of the trunk of the vehicle fell within the scope of the common law power. The appellant was arrested shortly after being removed from the vehicle that he was driving. A lawful search of his person disclosed a quantity of cannabis resin and a large quantity of money. He had been stopped in an area known to the officer for drug trafficking. In those circumstances, there was a reasonable prospect that the officer would find more drugs or narcotics in the vehicle. (at p. 200)

The accused was not immediately informed of his right to counsel and therefore the appeal was allowed and a new trial ordered.

CONCLUSION

This chapter has considered some of the essential elements that relate to the topic of arrest in Canadian law. Police officers have a significant power with regard to their capacity to relieve individual citizens of their liberty, through detention or arrest. We have examined the statutory authorities that guide the circumstances where an arrest, with or without a warrant, can be made under our existing criminal law, and we have seen that a wealth of case law exists where the courts have provided their assessment and judgment on specific instances where arrests have been made and people have been taken into custody.

The role of the Charter has been reviewed in the context of detention and arrest. Again, several cases have been summarized in order to assist the reader in seeing that there is a significant connection between the protections provided in the Charter and the application of police powers in this area of police activity.

QUESTIONS FOR CONSIDERATION AND DISCUSSION

1. What does the term "public interest" mean in the context of section 495(2)(d) of the *Criminal Code*?
2. What makes the arrest of a young offender different from the arrest of an adult?
3. Should the provisions for warrantless arrests be expanded or reduced? Given developments in technology, is it now feasible to require some form of warrant (e.g., telewarrant) in all instances?
4. Discuss the different uses of the terms "detain" and "arrest" in the *Criminal Code of Canada*.

CASES CITED

Brown v. Durham Regional Police Force (1999), 131 C.C.C. (3d) 1 (Ont. C.A.)

Koechlin v. Waugh and Hamilton (1957), 118 C.C.C. 24 (Ont. C.A.)

R. v. Chromiak (1979), 49 C.C.C. (2d) 257 (S.C.C.)

R. v. Clarkson (1986), 25 C.C.C. (3d) 207 (S.C.C.)

R. v. Dubois (1990), 54 C.C.C. (3d) 166 (Que. C.A.)

R. v. Feeney (1997), 115 C.C.C. (3d) 129 (S.C.C.)

R. v. Labine (1987), 49 M.V.R. 24 (B.C. Co. Ct.)

R. v. Ladouceur (1990), 56 C.C.C. (3d) 22 (S.C.C.)

R. v. Macooh (1993), 22 C.R. (4th) 70, 82 C.C.C. (3d) 481 (S.C.C.)

R. v. Manninen (1987), 58 C.R. (3d) 97 (S.C.C.)

R. v. Polashek (1999), 134 C.C.C. (3d) 187 (Ont. C.A.)

R. v. Simpson (1993), 79 C.C.C. (3d) 482 (Ont. C.A.)

R. v. Storrey (1990), 53 C.C.C. (3d) 316 (S.C.C.)

R. v. Therens (1985), 18 C.C.C. (3d) 481 (S.C.C.)

R. v. Tremblay (1987), 60 C.R. (3d) 59 (S.C.C.)

R. v. Vanstaceghem (1987), 58 C.R. (3d) 121 (Ont. C.A.)

R. v. Whitfield (1970), 1 C.C.C. 129 (S.C.C.)

REFERENCES

Canada. Canadian Committee on Corrections (1969). *Toward unity: criminal justice and corrections; report*. [Ottawa: Queen's Printer].

Canada. Criminal Law Review. Police Powers Project (1986). *Powers and procedures with respect to the investigation of criminal offences and the apprehension of criminal offenders: proposals with commentary*. Ottawa: Criminal Law Review.

Canada. Department of Justice (1971). *Manual respecting the authority and duties of peace officers in relation to arrest and pre-trial release and detention of accused persons*. Ottawa: Information Canada.

Law Reform Commission of Canada (1985). *Arrest*. Ottawa: The Commission. (Working paper; no. 41).

Manning, Morris (1983). *Rights, freedoms and the courts: a practical analysis of the Constitution Act, 1982*. Toronto: Emond-Montgomery.

Powell, Clay M. (1976). *Arrest and bail in Canada: a commentary on the Bail Reform Act, 1971, and the Criminal Law Amendment Act, 1975*. 2nd ed. Toronto: Butterworths.

Salhany, Roger E. (1997). *The police manual of arrest, seizure and interrogation*. 7th ed. Toronto: Carswell.

RELATED ACTIVITIES

- Examine relevant case law dealing with detention and arrest in Canada and other jurisdictions, particularly the United States.
- Follow stories pertaining to arrest carried in local or national media. Look for details, specifics, particular or special circumstances. What is the defence position? Have the police proceeded appropriately? What is the position of the Crown?
- Examine arrest statistics provided through the Centre for Justice Statistics (Statistics Canada). Make comparisons among different jurisdictions.

WEBLINKS

 http://www.lawsociety.mb.ca/help_rights_charged.htm This site outlines a citizen's rights under the Charter.

 http://www.opcc.bc.ca/Legal%20Reference%20Material/Entry%20and%20 Arrest%20in%20Dwelling%20Houses.html This paper explores some of the issues surrounding the Feeney judgment, and Parliament's response to Feeney, as reflected in Bill C-16.

 http://www.parlgc.ca/information/library/PRBpubs/8613-e.htm This page provides an oveview and explanation of the Young Offenders Act.

CHAPTER THREE

SEARCH AND SEIZURE
FOR CANADIAN POLICE

LEARNING OBJECTIVES

1. Identify sections of the *Criminal Code of Canada* dealing with the powers of search and seizure.
2. Provide a summary of key Charter decisions that impact on police powers of search and seizure in Canada.
3. Describe the procedures for obtaining a search warrant.
4. Identify when warrantless searches are permissible.
5. Describe the use of telewarrants in Canadian law.
6. Identify statutes, other than the *Criminal Code of Canada*, that include police powers of search.
7. Describe methods of updating statute and case law relevant to search and seizure in Canada.

INTRODUCTION

The police powers that relate to search and seizure in Canada are critical. They represent the capacity of the police to conduct highly intrusive searches on private premises and to secure bodily samples from individuals in the course of the investigation of criminal activity. Because of their capacity to invade upon the property and/or the person of Canadians, there is an important public interest in ensuring that these police powers are carefully controlled and circumscribed. This chapter will consider those powers in detail.

There is a connection with the *Canadian Charter of Rights and Freedoms* that will be explored, beyond the brief summary provided in Chapter 1. The significance of the search warrant will be discussed, as well as, the admissibility of illegally obtained evidence. Students will be provided with background on the general powers of search under the *Criminal Code of Canada*, and the procedures for obtaining a search warrant. The topic of telewarrants will also be addressed in some detail.

This chapter will also deal with the controversial use of video surveillance, tracking devices, and telephone recorders, which have all become available to police agencies in their efforts to control and prevent crime. Other areas that have a scientific or technological base will be reviewed, including the taking of bodily substances for DNA analysis and body impressions. The power of detention during a search will be addressed, as will be the seizure of goods and the duties that pertain to articles seized.

As the topic of search and seizure is both detailed and precise, the reader is provided with the following summary outline of this chapter's contents:

1. Preamble
2. *Charter of Rights and Freedoms*
3. Unreasonable search and seizure
4. The necessity of a search warrant
5. Admissibility of illegally obtained evidence
6. General powers of search under the *Criminal Code*
7. Procedure to obtain a search warrant
8. Warrantless searches (exigent circumstances)
9. Form of the warrant
10. Telewarrants
11. Video surveillance, tracking devices, and telephone recorders
12. Bodily substances for DNA analysis
13. Body impression warrants
14. Execution of the search warrant
15. Search of the media
16. Search of the person
17. Power of detention during a search
18. Seizure of goods
19. Duties after the articles are seized
20. Arrest of the person found in possession of stolen goods
21. Executing a warrant outside the territorial jurisdiction
22. Specific powers of search under the *Criminal Code*
23. Generally
    Prohibited and restricted weapons, firearms, or ammunition
    Gaming, betting, lotteries, and bawdy-houses
    Obscene publications, crime comics, and child pornography
    Blood samples
    Proceeds of crime

PREAMBLE

One issue that towers over others with regard to search and seizure is the accepted premise that citizens in Canada have a right to privacy and that the state is required to have compelling reasons before that right can be breached. To illustrate the strength and longevity of that right, it is helpful to refer to *Semayne's Case* (1604), 77 E.R. 194. In this case, from the 17th century, it was resolved:

> In all cases where the King is party, the sheriff may break the house, either to arrest or do other execution of the King's process, if he cannot otherwise enter. But he ought first to signify the cause of his coming and make request to open the doors.

This view respecting the sanctity of a person's home and belongings was confirmed in strong language in the case of *Entick v. Carrington* (1765), 95 E.R. 807 (K.B.), a famous British case that states the principle of self-incrimination and the search for evidence. The case dealt with trespass for breaking and entering. (See box below.)

ENTICK V. CARRINGTON (1765), 95 E.R. 807 (K.B.)

Facts:

Nathan Carrington and three others who were "messengers in ordinary to the King" entered the dwelling-house of John Entick on November 11, 1762. Over the course of four hours, without Entick's consent and against his will, these men searched for charts and pamphlets and broke open doors and locks. They had a warrant from the Secretary of State, which authorized them to make a "strict and diligent search" and they were specifically looking for material relating to a seditious paper: *The Monitor, or British Freeholder*.

Decision:

The court held that it was "contrary to the genius of the law of England" to issue such a warrant as it would jeopardize the liberty of the country. The court would not condone the ransacking of this man's home, including breaking into his drawers and boxes in a quest for evidence against him. On behalf of the court, the Lord Chief Justice stated:

ENTICK V. CARRINGTON (1765) continued

> . . . our law holds the property of every man so sacred, that no man may set his foot upon his neighbour's close without his leave; if he does he is a trespasser, though he does no damage at all; if he will tread upon his neighbour's ground, he must justify it by law . . . we can safely say there is no law in this country to justify the defendants in what they have done; if there was, it would destroy all the comforts of society; for papers are often the dearest property a man may have. (p. 817)

To search a person's house is an extreme measure that is supposed to be supported by a clear statutory provision permitting it. Police must be in possession of a warrant or some other specific authority before they can enter private premises and remain there against the owner's wishes.

Common law does recognize the implied right of anyone to come onto the owner's property. However, that right ends abruptly at the owner's front door. This is explained in more detail in the case of *R. v. Tricker* (1995), 96 C.C.C. (3d) 198 (Ont. C.A.).

In the case of *Hunter v. Southam Inc.* (1984), 41 C.R. (3d) 97 (S.C.C.) Mr. Justice Dickson provides some helpful background on the legal issue of protection against searches and seizures:

> Historically, the common law protections with regard to governmental searches and seizures were based on the right to enjoy property and were linked to the law of trespass. It was on this basis that in the great case of Entick v. Carrington (1765), 19 State Tr. 1029, 2 Wils. K.B. 275, 95 E.R. 807, the court refused to countenance a search purportedly authorized by the executive to discover evidence that might link the plaintiff to certain seditious libels. (p. 112)

The courts function not only to protect individuals from specific instances where police officers exceed their powers, but also to offer protection to the citizen from the abuses of power exercised by the government, even at the executive level.

CHARTER OF RIGHTS AND FREEDOMS

As we have already previewed in Chapter 1, section 8 of the Charter guarantees that:

> Everyone has the right to be secure against unreasonable search or seizure.

In Canada, all powers of search and seizure, statutory and common law, are subject to the constitutional protection found in the *Canadian Charter of Rights and Freedoms* under section 8.

It is thought that this section was inspired by the Fourth Amendment to the American *Constitution*. Prior to the Charter, it was still possible to admit evidence that had been obtained through an illegal search and seizure. The Charter substantially changed this approach.

Salhany (1997) provides valuable insight into the considerations that must be weighed by individual police officers when they are conducting a search:

> It is, therefore, important for a police officer to consider carefully the consequences of his conduct when conducting a search; otherwise he may discover that although he has discovered important and incriminating evidence against an accused, that evidence may be ruled inadmissible at trial. (p. 113)

In the case of *R. v. Rao* (1984), 46 O.R. (2d) 80, 40 C.R. (3d) 1, 12 C.C.C. (3d) 97 (Ont. C.A.) a person was charged with possession under the *Narcotic Control Act* (now repealed). Police officers found hash oil in Mr. Rao's office, upon entering without a search warrant. The trial judge held in this case that section 10(1)(a) of the *Narcotic Control Act*, which authorized a search without a warrant, was contrary to section 8 of the Charter. The Ontario Attorney General appealed this decision; however, that appeal was dismissed.

UNREASONABLE SEARCH AND SEIZURE

To elaborate on the meaning of a "reasonable" search, *Salhany* (1997) offers the following insight:

> A police officer must have reasonable cause to search a place. He only has reasonable cause when he is able to point to some information in his possession which would lead a reasonable person to conclude that his belief is probably true. (p. 113)

Take the example of a police officer approaching a motorist for the simple reason of drawing the driver's attention to a broken headlight. Should the officer discover that the person happens to be a drug trafficker, it would be unreasonable to search the vehicle. If there is some evidence of motion in the car when the officer approaches, that officer's suspicion will naturally increase; however, it still is not reasonable to conduct a search. If the driver denies making any motion, it then becomes possible to form a reasonable cause to search the vehicle.

Again, *Salhany* (1997) is helpful in the specifics of a search:

> What is important to recognize is that the officer must be able to justify his conduct. That conduct must be based upon reasonable cause. The test which the courts apply is whether a reasonable person, standing in the shoes of the police officer, would have believed that he had reasonable grounds. This is elaborated upon in Storrey, (1990), 75 C.R. (3d) 1 (S.C.C.). (p. 114)

The case of *R. v. Brezack*, [1949] O.R. 888, 9 C.R. 97 (Ont. C.A.) deals with an officer attempting to search an individual's mouth for drugs. The judgment in this case makes an important point about the difficult balance police officers are required to maintain in the execution of their duties:

> Constables have a task of great difficulty in their efforts to check the illegal traffic in opium and other prohibited drugs. Those who carry on the traffic are cunning, crafty and unscrupulous almost beyond belief. While, therefore, it is important that constables should be instructed that there are limits upon their right of search, including search of the person, they are not to be encumbered by technicalities in handling the situations with which they often have to deal in narcotic cases, which permit them little time for deliberation and require the stern exercise of such rights of search as they possess.

Unreasonable search and seizure is the theme of the case of *R. v. Mellenthin*, [1992] 3 S.C.R. 615. (See box below.)

R. V. MELLENTHIN, [1992] 3 S.C.R. 615

Facts:

As part of a check stop program, a police officer shone his flashlight into a car and saw an open gym bag that contained some glass vials. Upon further search, the officer found hash oil in those vials and some cannabis resin cigarettes.

Decision:

In explaining the court's decision in this matter, Justice Cory allows for the propriety of the flashlight check:

> There can be no quarrel with the visual inspection of the car by police officers. At night the inspection can only be carried out with the aid of a flashlight and it is necessary incidental to a check stop program carried out after dark. The inspection is essential for the protection of those on duty in check-stops. There have been more than enough incidents of violence to police officers when vehicles have been stopped. (pp. 623–624)

However, subsequent questions pertaining to the gym bag were found to be improper:

> Check stop programs result in the arbitrary detention of motorists. The programs are justified as a means aimed at reducing the terrible toll of death and injury so often occasioned by impaired drivers or by dangerous vehicles. The primary aim of the program is thus to check for sobriety, licences, ownership, insurance and the mechanical fitness of cars. The police use of check stops should not be extended beyond these aims. Random stop programs must not be turned into a means of conducting either an unfounded general inquisition or an unreasonable search. (p. 624)

R. V. MELLENTHIN (1992) continued

According to Cory, the search was unreasonable and, because of that determination, the evidence was inadmissible at trial:

> The appellant was detained at a check stop. While he was so detained, he was subjected to an unreasonable search. To admit the evidence obtained as a result of an unreasonable search of a motorist in a check stop would render the trial of the appellant unfair. Admitting such evidence would bring the administration of justice into disrepute. (p. 630)

What may have begun as a reasonable search could turn into one that is unreasonable through the actions of the police officer, as Manning (1983) notes:

> A difference occurring between the scope of the search authorized by the warrant and the scope of the search actually conducted may result in the search being unreasonable. While s. 445 of the Criminal Code permits the seizure of anything believed to have been obtained by or used in the commission of an offence in addition to those things specified in the warrant, the seizure of certain items may well be found to be unreasonable. The authority may not extend to items found after the listed articles are located and seized. The practice of the police in continuing to search for other matters after the goods listed are found may turn an otherwise reasonable search into an unreasonable one. Even though there is a search warrant in existence if the officer, for example, breaks down a door without good reason the search may be turned into an unreasonable search. (p. 295)

In the case of *R. v. Hynds* (1982), 8 W.C.B. 182 (Alta. Q.B.) an officer made a correction on a warrant that had an inaccurate street address. The officer felt that the occupants of the correct address recognized him and would have disposed of the evidence if he had sought a correction of the warrant. The court held that the evidence should not be excluded because the public would not be shocked by this conduct and it was based on an honest error.

The case of *Hunter v. Southam Inc.* (1984), 41 C.R. (3d) 97 (S.C.C.) is one of the most significant decisions to date on the Charter and has been extremely meaningful for Canadian jurisprudence. In his first pronouncement on the Charter, Chief Justice Dickson, on behalf of a unanimous court, advocated a strong, interventionist role for judges in exercising their Charter mandate. He also adopted a minimum standards approach to the protection against unreasonable search and seizure guaranteed by section 8.

In the case of *R. v. Collins*, [1987] 1 S.C.R. 265, 56 C.R. (3d) 193, 33 C.C.C. (3d) 1 (S.C.C.) the accused was in a pub with her husband. Police officers found heroin in her husband's car and she was holding a balloon that contained heroin. The majority of the Supreme Court of Canada held that a throat search was contrary to section 8 of the Charter if there were no reasonable grounds for the search.

Cloutier v. Langlois, [1990] 1 S.C.R. 158, 74 C.R. (3d) 316, 53 C.C.C. (3d) 257 (S.C.C.) deals with a charge of assault against two Montreal Urban Community police officers who arrested and searched a person for whom they had a warrant of committal for unpaid fines. Upon appeal the court arrived at a verdict of guilty. The arrest had been lawful, but the search was not and constituted an assault. It was held that a search requires the existence of reasonable grounds. At the Supreme Court of Canada, the appeal was allowed and the earlier acquittal of the officers was restored.

The issue of a "frisk-search" and its duration is considered in this judgment. Also, the court speaks to the issue of police discretion and its role in such procedures:

> In this regard a "frisk-search" is a relatively non-intrusive procedure: outside clothing is patted down to determine whether there is anything on the person of the arrested individual. Pockets may be examined, but the clothing is not removed and no physical force is applied. The duration of the search is only a few seconds. Though the search, if conducted, is in addition to the arrest, which generally entails a considerably longer and more sustained loss of freedom and dignity, a brief search does not constitute, in view of the objectives sought, a disproportionate interference with the freedom of persons lawfully arrested. There exists no less intrusive means of attaining these objectives.
>
> Furthermore, the fact that there existed a general directive in the police department to search an arrested suspect for any weapon or object potentially dangerous to the policemen has no bearing on this case, since the evidence of the police officers who conducted the search was that they exercised their independent discretion taking into account all the circumstances of this case.

The case of *R. v. Godoy* (1999), 131 C.C.C. (3d) 129 (S.C.C.) represents the first time a 911 call and the scope of police powers in responding were considered. At the Court of Appeal level, it was held that the police had a common law duty to investigate a 911 call and the authority to forcibly enter a dwelling to search for a caller. (See box below.)

R. V. GODOY (1999), 131 C.C.C. (3D) 129 (S.C.C.)

Facts:

Around 1:30 a.m. police arrived at a home to investigate a disconnected 911 call referred to as an "unknown trouble call." The appellant, Godoy, tried to prevent the police from entering his apartment. It was determined that the accused had hit his wife, and she had, in fact, called the police. Upon discovering that his wife had made this call, Godoy had disconnected the telephone.

Decision:

Speaking on behalf of the Supreme Court of Canada, Lamer C.J.C. makes the following assertion:

> I see no other use for an emergency response system if those persons who are dispatched to the scene cannot actually <u>respond</u> to the individual caller. I certainly cannot accept that the police should simply take the word of the person who answers the door that there is "no problem inside." (p. 138) [emphasis in original]

His decision goes on to elaborate a basis for the reasonableness of allowing police officers to enter premises under compelling circumstances:

> In summary, emergency response systems are established by municipalities to provide effective and immediate assistance to citizens in need. The 911 system is promoted as a system available to handle all manner of crises, including situations which have no criminal involvement whatsoever. When the police are dispatched to aid a 911 caller, they are carrying out their duty to protect life and prevent serious injury. This is especially true where the caller is disconnected and the nature of the emergency unknown. When a caller uses a 911 system, he or she has requested direct and immediate intervention and has the right to expect emergency services will arrive and locate the caller. The public interest in maintaining this system may result in limited intrusion in one's privacy interests while at home. This interference is authorized at common law as it falls within the scope of the police duty to protect life and safety and does not involve an unjustifiable use of the powers associated with this duty. (p. 141)

Again, in the case of *Hunter v. Southam Inc.* (1984), 41 C.R. (3d) 97 (S.C.C.), discussed briefly above, the issue of unreasonable search and seizure is addressed. (See box below.)

HUNTER V. SOUTHAM INC. (1984), 41 C.R. (3D) 97 (S.C.C.)

Facts:

Acting under section 10(1) of the *Combines Investigation Act*, the Director of Investigation and Research of the Combines Investigation Branch authorized representatives of the branch to enter the offices of a newspaper publishing company to examine and remove any book, paper record or other document that might afford evidence in connection with an inquiry into whether there should be charges laid for the offence of being party to a monopoly or that of lessening competition.

HUNTER V. SOUTHAM INC.   continued

Decision:

Authorization took place just prior to the proclamation of the Charter. A search was conducted that followed the provisions of the Charter. An application was made by Southam for an interim injunction. This application (i.e., that the search was unreasonable and a violation of section 8 of the Charter) was dismissed.

In the context of determining some definition for the term "unreasonable" Justice Dickson notes that:

> It is clear that the meaning of "unreasonable" cannot be determined by recourse to a dictionary, nor, for that matter, by reference to the rules of statutory construction. The task of expounding a constitution is crucially different from that of construing a statute. A statute defines present rights and obligations. It is easily enacted and as easily repealed. A constitution, by contrast, is drafted with an eye to the future. Its function is to provide a continuing framework for the legitimate exercise of governmental power and when joined by a bill or a charter of rights, for the unremitting protection of individual rights and liberties. Once enacted, its provisions cannot easily be repealed or amended. It must therefore be capable of growth and development over time to meet new social, political and historical realities often unimagined by its framers. (p. 110)

THE NECESSITY OF A SEARCH WARRANT

The *Canadian Charter of Rights and Freedoms* does not specifically require a peace officer to obtain a search warrant before conducting a search. However, in the case of *Hunter v. Southam Inc.* (1984), 41 C.R. (3d) 97 (S.C.C.) it was determined that a warrantless search may be presumed to be unreasonable. According to Chief Justice Dickson:

> A requirement of prior authorization, usually in the form of a valid warrant, has been a consistent prerequisite for a valid search and seizure both at common law and under most statutes. Such a requirement puts the onus on the State to demonstrate the superiority of its interests to that of the individual. As such it accords with the apparent intention of the Charter to prefer, where feasible, the right of the individual to be free from state interference to the interests of the State in advancing its purposes through such interference. (p. 109)

The issuance of a search warrant does not give a police officer a free hand with regard to any subsequent search. The officer is only permitted to search for those objects, and in those areas, outlined in the warrant document itself.

A search warrant may be determined to be defective or may have been improperly issued. The search then becomes a warrantless one. Evidence gained in this instance does not automatically become excluded under section 24(2) of the Charter, which was discussed in Chapter 1.

An exceedingly detailed and important case is *R. v. Church of Scientology* (1987), 31 C.C.C. (3d) 449 (Ont. C.A.) leave to appeal to S.C.C. refused (1987), 82 N.R. 393n (S.C.C.). Here, the Church of Scientology brought an application to quash a search warrant executed at their offices in Toronto. This represents an extensively detailed case with regard to search warrants. (See box below.)

R. V. CHURCH OF SCIENTOLOGY (1987), 31 C.C.C. (3D) 449 (ONT. C.A.) LEAVE TO APPEAL TO S.C.C. REFUSED (1987), 82 N.R. 393N (S.C.C.)

Facts:

In 1983 a detective sergeant with the OPP, Anti-Rackets Branch, swore an information before a justice to obtain a search warrant to search the premises of the Church. The information was more than one thousand pages in length. Chief Judge Hayes signed the warrant, which resulted in a massive, coordinated police search.

One hundred and twenty-nine OPP officers attended the search. Thirty officers conducted the search and they seized about 850 boxes of material, representing more than 39 000 files and books or approximately 2 000 000 documents, statements, and tapes. The sheer quantity of documentary evidence presented an enormous challenge to the court.

Decision:

In the court's judgment:

> Police work should not be frustrated by the meticulous examination of facts and law that is appropriate to a trial process. It may or may not be that the informant has honestly made mistakes in the preparation of this information that supports the warrant. There may be serious questions of law as to whether what is asserted amounts to a criminal offence. (p. 475)

Also:

> . . . it is most difficult to describe business records, books of account and financial statements generally with the same degree of particularity as one can describe such things as a motor vehicle, a book, a specific business agreement or a particular letter. The latter items can be identified by a model number, a licence number or by reference to a specific title or date of and parties to an agreement or letter. In such cases it is a simple matter to give exact descriptions and there is no reason for an officer executing a warrant to be given or to exercise any discretion. Where, however, by the very nature of the things to be searched for it is not possible to describe them with precision or great particularity, it is inevitable that [the officer must] . . . exercise some discretion in determining whether things found on the premises fall within the description of things or classes of things described in the warrant. (pp. 515–16)

In the case of *R. v. Ferguson* (1990), 1 C.R. (4th) 53 (Ont. C.A.) a warrantless strip search conducted at the side of the road produced evidence that was excluded from trial because it was in conflict with section 8 of the Charter and was highly intrusive.

The judgment of the court included the following observation with respect to "reasonable and probable grounds":

> Their belief in the existence of reasonable and probable grounds, if indeed they had such a belief, was not in our view a reasonable one. Their strong suspicion, even though it turned out to be right, is simply not enough, and did not justify their conduct. In our view, the admission of the evidence discovered as a result of the search, notwithstanding that it was real evidence, would bring the administration of justice into disrepute. (p. 56)

As a result, the appeal was allowed, and an acquittal was substituted.

ADMISSIBILITY OF ILLEGALLY OBTAINED EVIDENCE

Under section 24 of the Charter, evidence that has been illegally obtained will only be excluded if it is established that, taking all of the circumstances into account, its admission would bring the administration of justice into disrepute (see *Chapin* (1983), 7 C.C.C. (3d) 538 at 541 (Ont. C.A.)). The onus is on the accused to show that the administration of justice would be brought into disrepute.

Salhany makes the following comments on the case of *Collins* (1987), 56 C.R. (3d) 193, 33 C.C.C. (3d) 1, [1987] 1 S.C.R. 265:

> In the Collins case, the Court said that there are three factors which a trial judge must consider in determining whether the administration of justice has been brought into disrepute. The first factor (already discussed) is whether the admission of evidence will affect the fairness of the trial. The second group of factors concern the seriousness of the violation. Relevant to this group will be whether the violation was committed in good faith, whether it was inadvertent or of a merely technical nature, whether it was motivated by urgency or to prevent loss of evidence, and whether the evidence could have been obtained without a Charter violation. (p. 119)

In *Stillman*, [1997] 1 S.C.R. 607, the court determined that hair samples, buccal swabs, and dental impressions taken from the accused against his wishes, and by means of threats, should be excluded because they were acquired contrary to the section 7 Charter rights of the accused. However, in this case, a tissue discarded by the accused, which happened to contain DNA material that could be analyzed, was seen as admissible.

GENERAL POWERS OF SEARCH UNDER THE *CRIMINAL CODE*

A justice of the peace can only issue a warrant authorizing a daytime search of premises for stolen property where information has been sworn before him alleging a theft has been committed. This is established under common law. Section 487(1) of the *Criminal*

Code extends that authority. As a result, a justice can issue a search warrant for stolen goods, and also for:

1. Anything relating to an offence which has been or is suspected to have been committed against the *Criminal Code* or any other Act of Parliament;
2. Anything that she or he reasonably believes will be evidence of an offence or will reveal the whereabouts of a person who is believed to have committed an offence against the *Criminal Code* or any other Act of Parliament; or
3. Anything that is reasonably believed to be intended to be used to commit an offence against a person for which the offender may be arrested without a warrant.

This section authorizes searches in a building, a receptacle, or a place. Courts have construed these provisions very strictly. See *Re Worrall* (1965), 44 C.R. 151 (Ont. C.A.) for a closer examination of the meaning of a "building" or "receptacle" and consideration about the meaning of "place." In *Re Laporte* (1972), 18 C.R.N.S. 357 (Que. Q.B.) it was viewed that the body was not a "place."

Section 256(1) of the *Criminal Code* (enacted in 1985) gives statutory authority for the granting of a search warrant for extracting blood samples from a person unable to give consent. The justice issuing such a warrant must be satisfied that there are reasonable grounds to believe that the person has, within the last two hours, committed the offence of impaired driving and was in an accident resulting in death or bodily harm to any person.

Confusion as to what constitutes a "place" can be seen in *Stevens* (1983), 7 C.C.C. (3d) 260 (N.S.C.A.) and *Rao* (1984), 12 C.C.C. (3d) 97 (Ont. C.A.). Places here were seen as fixed locations, such as offices, shops, gardens, as well as, vehicles, vessels, and aircraft. Not included were public streets and public places.

Salhany has emphasized a point that differentiates between searches that seek out direct evidence of crime and those that "fish" for evidence:

> The purpose of a search warrant is to assist the administration of justice by enabling the police to enter premises for the purposes of locating items that will provide evidence that a crime has been committed. It must not be used by police officers to enable them to make observations regarding how things are used so that evidence of a crime may be obtained. (p. 124)

PROCEDURE TO OBTAIN A SEARCH WARRANT

An information must be sworn upon oath before a justice of the peace before a search warrant may be issued. The warrant must be in Form 1 (see Figure 3.1, below) found in the *Criminal Code*.

FIGURE 3.1

FORM 1 – INFORMATION TO OBTAIN A SEARCH WARRANT
(Section 487)

Canada,

Province of ....,
(territorial division).

This is the information of A.B., of ...., in the said *(territorial division)*, *(occupation)*, hereinafter called the informant, taken before me.

The informant says that *(describe things to be searched for and offence in respect of which search is to be made)*, and that he believes on reasonable grounds that the said things, or some part of them, are in the *(dwelling-house, etc.)* of C.D., of ...., in the said *(territorial division)*. *(Here add the grounds of belief, whatever they may be)*.

Wherefore the informant prays that a search warrant may be granted to search the said (dwelling-house, etc.) for the said things.

Sworn before me
this .... day of
........., A.D. ...., ...
at ....   (Signature of Informant)

....
A Justice of the Peace in and
for

In addition to the name, address, and occupation of the informant, the following items must be included in Form 1:

- *Description of the articles to be searched for*—there should be a full description of the items to be searched for, if this is possible. This level of detail will assist the police officer in satisfying the justice of the peace that the goods described do exist, that they are in the possession of the accused and that they are related to a specific offence which has been or is suspected to have been committed against the *Criminal Code* or any other Act of Parliament. The justice is the person who must be satisfied, not the officer. An additional reason for a full level of detail is to ensure that the items will be readily identifiable when the search is conducted. Courts typically frown on search warrants that offer only vague descriptions of goods or items, and may be seen as leading to "fishing expeditions" by the police when searching premises. On the other hand, the courts realize that some cases are exceedingly complex and the level of detail desired for the search warrant may not be available. A good example of this is the case *Church of Scientology* (1987), 31 C.C.C. (3d) 449 (Ont. C.A.) leave to appeal to the S.C.C. refused (1987), 82 N.R. 393n. The goods subject to seizure must be personal property, not real property, so in one instance it was deemed that doors from a house were part of real property and so could not be seized;

- *Statement of the offence involved*—the information should contain details of the offence for which the goods are being sought. This type of information should also be found in the warrant. Reasons for this are provided in *Solloway Mills & Co.* (1930), 53 C.C.C. 261 (Alta. C.A.), in particular to inform the owner of the premises of the reason for the search and of his or her right to see the warrant; and

- *Grounds of belief*—the informant must swear that he or she has reasonable grounds for believing that specific articles are in the building to be searched, and articulate the grounds for that belief. This is to assist the justice in making a determination about the facts and to decide whether or not to exercise discretion in the issuance of the warrant to search.

No search warrant should be granted lightly. The justice is carrying out a judicial act and must be conscious of the significance of these decisions. The concept of acting judicially is dealt with in the case of *Moran* (1982), 65 C.C.C. (2d) 129 (S.C.C.).

There are instances where the person who has provided information should not be identified. The wording for such an information is as follows:

> . . . as he has been informed by a person (or persons) whose name cannot be disclosed for reasons of public policy . . .

See the case of *Swarts* (1916), 27 C.C.C. 90 (Ont. H.C.) for an explanation of this area.

WARRANTLESS SEARCHES
(EXIGENT CIRCUMSTANCES)

The case of *R. v. Rao* (1984), 12 C.C.C. (3d) 97 (Ont. C.A.) is an extremely important case dealing with warrantless search and seizure.

Martin J.A. has noted:

The Law Reform Commission of Canada has stated, the warrantless powers currently given to peace officers in this country with respect to narcotics and drugs exceed those available in most of the other countries with a common law tradition: Working Paper 30, Police Powers—Search and Seizure in Criminal Law Enforcement, at p. 282. This statement is correct in so far as it relates to the power conferred on peace officers in Canada to search private premises without a warrant. The power of the police to search persons without a warrant who are suspected to possess narcotic drugs is, however, as previously pointed out, more limited in Canada than in some other Commonwealth countries. (pp. 120–121)

The case of *Cloutier v. Langlois* (1990), 74 C.R. (3d) 316 (S.C.C.) deals with warrantless searches. (See box below.)

CLOUTIER V. LANGLOIS (1990), 74 C.R. (3D) 316 (S.C.C.)

Facts:

Members of the Montreal Urban Community Police stopped a motor vehicle that was seen violating a municipal by-law. Specifically, the driver made a right turn from a centre lane, directly in front of the officers. It was discovered that there was a warrant of committal for unpaid fines against the registered owner. When the officers stopped the driver for the violation and the unpaid fines, he reacted in a decidedly unpleasant, highly agitated manner. Throughout his interaction with the police, Mr. Cloutier was verbally abusive. Out of concern for their safety, the officers conducted a frisk-search. Eventually, it was determined that Mr. Cloutier was a lawyer practising in Montreal. In this case, the court allowed the appeal and the original acquittal was restored.

Decision:

This case contains a good review of the origin and evolution of the common law rule pertaining to search.

Justice L'Heureux-Dubé has ruled:

. . . the ultimate purpose of criminal proceedings is to convict those found guilty beyond a reasonable doubt. Our system of criminal justice is based on the punishment of conduct that is contrary to the fundamental values of society, as statutorily enshrined in the Criminal Code and similar statutes. That is its primary purpose. The system depends for its legitimacy on the safe and effective performance of this function by the police. In the context of an arrest, these requirements entail at least two primary considerations. First, the process of arrest must be capable of ensuring that those arrested will come before the court. An individual who is arrested should not be able to evade the police before he is released in accordance with the rules of criminal procedure, otherwise the administration of justice will be brought into disrepute. In light of this consideration, a search of the accused for weapons or other dangerous articles is necessary as an elementary precaution to preclude the possibility of their use against the police, the nearby public or the accused himself. Incidents of this kind are not unknown. Farther, the process of arrest must ensure that evidence found on the accused and in his immediate surroundings is preserved. The effectiveness of the system depends in part on the ability of peace officers to collect evidence that can be used in establishing the guilt of a suspect beyond a reasonable doubt.

However, while the common law gives the police the powers necessary for the effective and safe application of the law, it does not allow them to place themselves above the law and use their powers to intimidate citizens. This is where the protection of privacy and of individual freedoms becomes very important. (pp. 361–362)

The case of *R. v. Stillman*, [1997] 1 S.C.R. 607 examines significant issues with regard to the power of search incidental to arrest. (See box below.)

R. V. STILLMAN, [1997] 1 S.C.R. 607

Facts:

The accused was a seventeen-year-old male, arrested for the murder of his 14-year-old girlfriend. He was the last person to see her and arrived home in a dishevelled state with a cut over one eye. The victim had been sexually assaulted, with human bite marks on her abdomen.

The lawyer for the accused notified the police that he was not consenting to provide bodily samples, including hair or teeth imprints. When the lawyer left, the police did take hair samples and the accused was compelled to remove his own pubic hair. Plasticine teeth imprints were also taken. Intensive interviewing took place and the accused was allowed to call his lawyer. In the washroom, he blew his nose and threw away the tissue.

R. V. STILLMAN (1997)   **continued**

This was seized by the police for DNA testing. He was released and re-arrested a number of months later. A dentist took new teeth impressions without his permission and more hair samples were taken along with saliva samples and buccal swabs.

Decision:

It was held that the bodily samples had been obtained in violation of section 8 of the Charter. His arrest was lawful because the police had reasonable and probable grounds.

The decision, by Justice Cory, includes the following statement on the limitation to the common law power of search incidental to arrest:

> However, the common law power of search incidental to arrest does not extend beyond the purpose of protecting the arresting officer from armed or dangerous suspects or of preserving evidence that may go out of existence or be otherwise lost. The search conducted in this case went far beyond the typical "frisk" search which usually accompanies an arrest. (p. 610)

It was noted that the conduct of the police was particularly bad in light of the fact that this was a young offender.

Justice Cory notes at page 612:

> The accused's bodily samples and impressions existed as "real" evidence, but the police, by their words and actions, compelled the accused to provide evidence from his body. This evidence constituted conscriptive evidence. The impugned evidence would not have been discovered had it not been for the conscription of the accused in violation of his Charter rights and no independent source existed by which the police could have obtained the evidence.

And at page 613:

> . . . the fact that the police rode roughshod over a young offender's refusal to provide his bodily samples would certainly shock the conscience of all fair minded members of the community.

Tissue and mucus was *not* excluded in spite of being taken surreptitiously.

Justice Cory notes at page 640:

> It is certainly significant that Parliament has recently amended the Criminal Code through the addition of s. 487.05, so as to create a warrant procedure for the seizure of certain bodily substances for the purposes of DNA testing. This suggests that Parliament has recognized the intrusive nature of seizing bodily samples.

And at page 642:

> No matter what may be the pressing temptation to obtain evidence from a person the police believe to be guilty of a terrible crime, and no matter what the past frustrations to their investigations, the police authority to search as an incident to arrest should not be exceeded. Any other conclusion could all too easily lead to police abuses in the name of the good of society as

perceived by the officers . . . The treatment meted out by agents of the state to even the least deserving individual will often indicate the treatment that all citizens of the state may ultimately expect.

Also, at page 658:

Canadians think of their bodies as the outward manifestations of themselves. It is considered to be uniquely important and uniquely theirs. Any invasion of the body is an invasion of the particular person. Indeed, it is the ultimate invasion of personal dignity and privacy. No doubt this approach was the basis for the assault and sexual assault provisions. The body was very rightly seen to be worthy of protection by means of criminal sanctions . . . The concept of fairness requires that searches carried out in the course of police investigations recognize the importance of the body.

And at page 660:

If there is not respect for the dignity of the individual and the integrity of the body then it is but a very short step to justifying the exercise of any physical force by police if it is undertaken with the aim of solving crimes.

To remedy the dilemma posed by the *Feeney* case (discussed in detail in Chapter 2), on May 14, 1997, the Parliament of Canada proclaimed the *Criminal Law Improvement Act* into force. Section 487.11 of the amendment reads as follows:

A peace officer, or a public official who has been appointed or designated to administer or enforce any federal or provincial law and whose duties include the enforcement of this or any other Acts of Parliament, may, in the course of his or her duties, exercise any of the powers described in subsection 487(1) or 492.1(1) without a warrant if the conditions for obtaining a warrant exist but by reason of exigent circumstances it would be impracticable to obtain a warrant.

Such circumstances could presumably include imminent danger of loss, removal, destruction, or disappearance of the evidence if the search and/or seizure are delayed.

The question of exigent circumstances was raised in the case of *Feeney* (1997), 115 C.C.C. (3d) 129 (S.C.C.). The court, however, did not address the question directly, and it is not certain whether section 487.11 will stand a constitutional challenge. Presently, however, police officers may use its provisions when they see fit.

FORM OF THE WARRANT

Form 5 (see Figure 3.2, below), contained in the *Criminal Code*, outlines the contents of the warrant itself and the details it must contain.

FIGURE 3.2

FORM 5 – WARRANT TO SEARCH
(Section 487)

Canada,

Province of ....,
(territorial division)

To the peace officers in the said (territorial division) or to the (named public officers):

Whereas it appears on the oath of A.B., of .... that there are reasonable grounds for believing that *(describe things to be searched for and offence in respect of which search is to be made)* are in .... at .... , hereinafter called the premises;

This is, therefore, to authorize and require you between the hours of *(as the justice may direct)* to enter into the said premises and to search for the said things and to bring them before me or some other justice.

Dated this .... day of .... A.D. .... , at ....

....
A Justice of the Peace in and for
....

1999, c. 5, s. 4

Officers do not need to be named on the warrant. This point is verified in the case of *Haley* (1986), 27 C.C.C. (3d) 454 (Ont. C.A.). However, the warrant will also contain a specific authorization and direction that the search be conducted between specific hours of the day. Section 488 of the *Criminal Code* makes it necessary to execute the warrant "by day" unless there are reasonable grounds for the justice to authorize execution of the warrant by night and she or he explicitly authorizes this approach. "Day" is defined in the *Criminal Code* as that period between 6:00 a.m.

and 9:00 p.m. Courts typically frown on nighttime searches, as is apparent from a reading of the case of *R. v. Posternak* (1929), 51 C.C.C. 426 (Alta. C.A.).

TELEWARRANTS

Provisions were introduced to amend the *Criminal Code*, on December 2, 1985, to authorize peace officers to obtain search warrants from a judicial officer by telephone, or other telecommunication, where the officer believes that an indictable offence has been committed. The resulting provisions are set out in section 487.1 of the *Criminal Code*. They have come to be referred to as "telewarrants" and they are obtained as follows:

> A peace officer who believes that it is impracticable to appear personally before a justice may apply for a search warrant by telephone or other means of communication (s. 487.1(1)).

Application is made by submitting an information on oath by telephone or other telecommunication to the justice who shall record it verbatim. The justice will then certify the record of the information as to time, date, and contents and file it with the clerk of the territorial division in which the warrant is intended for execution (section 487.1(2)).

The information shall include:

1. a statement of the circumstances making it impracticable to appear before a justice;
2. a statement of the indictable offence alleged, the place to be searched, and the items to be seized;
3. the officer's grounds for believing that the items will be found in the place to be searched; and
4. a statement as to any prior application for a warrant pertaining to the same matter (section 487.1(4)).

If the justice is satisfied that the requirements in section 487.1(3) have been fulfilled, that there are reasonable grounds to dispense with an information in writing, and that there are reasonable grounds for the warrant, he or she will issue it (section 487.1(5)).

The justice must complete and sign the warrant in Form 5.1, noting the time, date, and place of issuance; the peace officer must complete in duplicate a facsimile of the warrant, noting the name of the issuing justice and the time, date, and place of issuance; the justice must then file the warrant as soon as practicable with the clerk of the court for the territorial division in which it is to be executed (section 487.1(6)).

The officer executing the warrant is required to give a facsimile to anyone in control of the premises as soon as practicable (section 487.1(7)). If the premises are unoccupied, the facsimile must be affixed in a prominent place inside the premises (section 487.1(8)).

The officer who receives the warrant is required to file a written report with the clerk of the court in the territorial division where it is intended for execution, within seven days after it is executed. The report must contain the following:

1. a statement of time and date of execution; if not executed, why it was not;
2. a list of the items seized and where they are being held; and
3. a list of the items seized in addition to those mentioned in the warrant, where they are being held, and the reason for their seizure (section 487.1(9)).

The clerk is then required to bring the report, information, and telewarrant before a justice to be dealt with as if the officer had obtained the warrant personally (section 487.1(10)).

The absence of an information on oath, transcribed and certified by the justice as to time, date, and contents, or the original warrant signed by the justice with notation on its face of the time, date, and place of issuance, is, in the absence of evidence to the contrary, proof that the search and seizure was not properly authorized where it is material in any proceeding before the court (section 487.1(11)).

VIDEO SURVEILLANCE, TRACKING DEVICES, AND TELEPHONE RECORDERS

Parliament proclaimed several search and seizure provisions dealing with video surveillance, tracking devices, and telephone recorders, designed to help police officers (*Criminal Code*, S.C. 1993, c. 40, s.15). Section 487.01 now authorizes video surveillance where it has been granted by a judicial warrant. A judge of the provincial court, or the superior court of criminal jurisdiction, has the authority to issue a warrant in writing authorizing a peace officer to use any device, investigative technique, or procedure that would otherwise constitute an unreasonable search and seizure of a person so long as it does not interfere with her or his bodily integrity or property if:

1. the judge is satisfied by information on oath in writing that there are reasonable grounds to believe that an offence has been or will be committed against an Act of Parliament and that information will be obtained by such device, etc.;
2. the judge is satisfied that it is in the best interests of the administration of justice to issue the warrant; and
3. there is no other statutory provision authorizing its use.

A police officer appears before a judge to obtain a search warrant.

Included in the warrant will be any terms or conditions that the justice considers appropriate in order to ensure that the search and seizure is reasonable. The person who is to be placed under video surveillance has a reasonable expectation of privacy and the warrant must contain terms and conditions that will ensure that such privacy is respected.

BODILY SUBSTANCES FOR DNA ANALYSIS

The *Criminal Code* was amended in 1995, along with the *Young Offenders Act*, to give provincial court judges jurisdiction to issue a warrant authorizing a peace officer or another person acting under a peace officer's direction, to obtain samples of bodily substances by means of special investigative procedures for forensic DNA analysis from a person who is reasonably believed to have been a party to certain designated *Criminal Code* offences. Those designated offences are set out in section 487.04 of the Code and include crimes of violence such as murder and manslaughter, sexual offences, and arson.

An application for such a warrant may be made *ex parte*; however, it must be by information on oath, as per section 487.05(1). The informant is required to state that he or she has reasonable grounds to believe:

(a) that a designated offence has been committed;
(b) that a bodily substance has been found
    (i) at the place where the offence was committed,
    (ii) on or within the body of the victim of the offence,
    (iii) on anything worn or carried by the victim at the time when the offence was committed, or

(iv) on or within the body of any person or thing or at any place associated with the commission of the offence,

(c) that a person was a party to the offence, and

(d) that forensic DNA analysis of a bodily substance from the person will provide evidence about whether the bodily substance referred to in paragraph (b) was from that person.

The judge who issues the warrant must be satisfied that it is in the best interests of the administration of justice that this warrant be issued (section 487.05(1)). Prior to issuing the warrant, the judge must have regard to all of the relevant issues, including the following:

(a) the nature of the designated offence and the circumstances of its commission, and

(b) whether there is
   (i) a peace officer who is able, by virtue of training or experience, to obtain a bodily substance from the person, by means of an investigative procedure described in subsection 487.06(1), or
   (ii) another person who is able, by virtue of training or experience, to obtain under the direction of a peace officer, a bodily substance from the person, by means of such an investigative procedure, s. 487.05(2).

Three investigative procedures are permitted to collect the bodily substance (section 487.06(1)), including:

(a) the plucking of individual hairs from the person, including the root sheath;

(b) the taking of buccal swabs by swabbing the lips, tongue and inside cheeks of the mouth to collect epithelial cells; or

(c) the taking of blood by pricking the skin surface with a sterile lancet.

Specific terms and conditions with regard to ensuring that the seizure of bodily substances is reasonable shall be included by the judge (section 487.06(2)).

Prior to executing such a warrant, the peace officer is required to inform the person of the following:

(a) the contents of the warrant;

(b) the nature of the investigative procedure to obtain the bodily substance;

(c) the purpose of obtaining the bodily substance;

(d) the possibility that the results of forensic DNA analysis may be used in evidence;

(e) the authority of the peace officer and any other person under his direction to use as much force as is necessary to execute the warrant; and

(f) in the case of a young person, his or her rights under s. 487.07(4): s. 487.07(1).

The person who is the subject of the warrant may be detained for a period that is reasonable in order to obtain a bodily substance from that person. The person may also be required to accompany the officer for this purpose, as per section 487.07(2).

The bodily substance and the forensic results are to be used only for the investigation of the offence designated on the warrant. Both the substance and the results must be destroyed "forthwith" if the involvement of the suspect in the commission of the offence has been disproved (section 487.09(1)(a)). Both must also be destroyed if the accused is acquitted of the offence (section 487.09(1)(b)). Also, both must be destroyed if the suspect is discharged following a preliminary inquiry, or the charge is dismissed, withdrawn, or stayed, unless a new information is laid or an indictment is preferred within a year (section 487.09(1)(c)).

However, a provincial court judge may direct that the bodily substance or analysis not be destroyed if he or she is satisfied that these items might be reasonably required in an investigation or prosecution of that person for another designated offence, or of another person for that designated offence, or any other offence in respect of the same transaction (section 487.09(2)).

BODY IMPRESSION WARRANTS

Parliament proclaimed a *Criminal Law Amendment Act* in May 1997 that added section 487.091 to the *Criminal Code*. The section gives a justice of the peace jurisdiction to issue a warrant authorizing a peace officer to take or cause to be taken a "handprint, fingerprint, footprint, foot impression, teeth impression or other print or impression of the body or any part of the body." Terms and conditions may be added to such a warrant to ensure that the search and seizure are reasonable. Application by telewarrant may be made by an officer where it is impracticable to appear before an issuing judge or justice.

EXECUTION OF THE SEARCH WARRANT

A search warrant may be executed on any day of the week, including a holiday. The officer must carry the warrant and produce it if she or he is requested to do so (*Criminal Code*, section 29(1)). A formal demand should be made by the officer to the person in charge of the premises to be admitted. Upon refusal, the officer may break into the premises to execute the warrant (see *Wah Kie v. Cuddy (No. 2)* (1914), 23 C.C.C. 383 (Alta. C.A.)). Prior to breaking into cabinets or other locked areas, the officer should first demand production of the keys that will afford entry to those places.

No prior announcement is required for police officers executing a warrant granted under section 11 of the *Controlled Drugs and Substances Act*. In the case of *Gimson* (1990), 54 C.C.C. (3d) 232 (Ont. C.A.), affirmed 69 C.C.C. (3d) 552n (S.C.C.) the court held that sections 10 and 12 and 14 of the *Narcotic Control Act* (now repealed, and which preceded the *Controlled Drugs and Substances Act*) gave to the police special powers of search that constitute a comprehensive set of rules, replacing the older common law rules relating to the execution of search warrants. The element of surprise is clearly seen as being important to the police when they have reason to believe that they are dealing with drugs. There was a warning by the court, though, that the search must be carried out in a reasonable manner to ensure that it complies with section 8 of the Charter.

It is typically worthwhile for the officers to plan the role of each officer involved in a search. This would avoid the search as seeming to be random and avoid any judicial criticism.

SEARCH OF THE MEDIA

Salhany (1997) presents the following summary with respect to a search of the media:

> When it comes to the media, the courts believe that serious interference with the operation of a media organization must be counterbalanced by the necessity to obtain information. This necessity can only be demonstrated where there is no other reasonable alternative source or, if there is, it has been pursued unsuccessfully . . . (p. 144)

This issue was examined in the case of *Radio Canada c. Nouveau-Brunswick (P.G.)*(1992), 9 C.R. (4th) 192 (S.C.C.).

SEARCH OF THE PERSON

An officer who is conducting a search of a place has a right to search a person for an offence against the *Controlled Drugs and Substances Act*. However, that officer does not have a right to arbitrarily search anyone found in a public place or on the premises where a search is being conducted, except in relation to a weapons offence.

Section 11 of the *Controlled Drugs and Substances Act* allows an officer who is executing a search warrant to search any person found in the place referred to in the warrant. The officer must have a reasonable belief that the person being searched is in possession of the narcotic or prohibited drug. This matter is considered in the case of *Debot* (1986), 54 C.R. (3d) 120, 30 C.C.C. (3d) 207 (Ont. C.A.). The officer does not have a right to search anyone and everyone found in the place. He or she must establish some basis for a reasonable belief that the person is carrying a narcotic or drug.

Police officer conducting a search of a suspect.

Credit: Mike Weaver, Media Relations, Kingston Police

POWER OF DETENTION DURING A SEARCH

The question of what to do with people found on premises being searched for drugs came up in the case of *Levitz v. Ryan* (1972), 9 C.C.C. (2d) 182 (Ont. C.A.). Here the RCMP was executing a warrant and the officers felt that the occupants were under arrest for the period of the search. A civil action was brought against the officers for assault, false imprisonment, and conspiracy to injure. This action was dismissed. The judge pointed out that there was no Canadian jurisprudence on the matter of the right of police officers to "freeze" the premises while conducting a search. The purpose of the search might have been thwarted had one of the occupants been permitted to leave or make a telephone call. However, the detention of the occupants was not authorized automatically in every case; officers still require reasonable grounds for believing that such a detention is necessary.

SEIZURE OF GOODS

An officer who is executing a warrant is entitled to seize any article named in the warrant. The officer is also entitled to seize anything that she or he believes, on reasonable grounds, has been obtained by or has been used in the commission of an offence, or will afford evidence in respect of any offence against the *Criminal Code* or any other Act of Parliament. But the officer must be cautious that she or he does not take anything that could not possibly provide evidence in a criminal charge.

An officer may seize any explosive substance that the officer suspects may be used for an unlawful purpose (*Criminal Code*, section 492(1)).

Section 489(2), which was added to the *Criminal Code* in May 1997, authorizes the police to seize anything without a warrant that is reasonably believed to have been obtained or used in the commission of a federal offence or will afford evidence of such an offence. This still requires that which is seized to be in "plain view" of the officer.

DUTIES AFTER THE ARTICLES ARE SEIZED

Any goods seized as a result of executing a search warrant must be brought before the justice who issued the warrant, under section 487(1) of the *Criminal Code*, or to some other justice in the same territorial division. Alternatively, the officer may make a report to the justice in accordance with section 489.1 of the *Criminal Code* instead of making a physical return. The report must follow Form 5.2 with variations to suit the case at hand and must comply, in the case of telewarrants, with section 487.1(9).

Section 489.1 also requires a police officer to return seized property to a person lawfully entitled to possession of it if the officer is satisfied that there is no dispute regarding ownership, and if the items are not required for the purposes of any further investigation, a preliminary inquiry, the trial, or some other proceeding. The officer is then required to report what has been done to the justice.

Section 490 of the *Criminal Code* has been amended to require that the justice of the peace return goods seized to the lawful owner, unless the prosecutor can satisfy him or her that they are required for some proceeding.

ARREST OF THE PERSON FOUND IN POSSESSION OF STOLEN GOODS

Whether or not a person found on the premises being searched should be arrested will depend on whether the officer concludes that she or he is justified in doing this under her or his general powers of arrest. Unless the officer conducting the search believes on reasonable grounds that the person has committed or is about to commit an indictable offence, or finds that person committing a criminal offence in relation to the goods found, the officer is not justified in arresting that person.

EXECUTING A WARRANT OUTSIDE THE TERRITORIAL JURISDICTION

Should it occur that the premises to be searched are outside the issuing justice's jurisdiction, section 487(2) of the *Criminal Code* permits the warrant to be taken to a justice who has jurisdiction to endorse the warrant. This is referred to as "backing the warrant" and is carried out in accordance with Form 28. Once the warrant has been endorsed then it may be executed, not only by the officer or officers to whom it was originally directed, but also by any other peace officer in that jurisdiction (section 487(4)).

SPECIFIC POWERS OF SEARCH UNDER THE *CRIMINAL CODE*

The *Criminal Code* authorizes specific powers of search and seizure for certain identified offences. A power to search premises will exist only if the relevant section grants that power.

In the case of *R. v. Colet* (1981), 19 C.R. (3d) 84 (S.C.C.) the police held a warrant that authorized them to seize any firearms or offensive weapons in the possession of the accused pursuant to the former section 105(1) of the *Criminal Code*. The officers went to the accused's house and he refused to let them in and attacked the police. At the accused's subsequent trial he was acquitted and the court emphasized that there was a difference between the right to enter a person's home "to seize" and the right "to search." They held that in this instance the police could seize but they did not have the right to search.

GENERALLY

PROHIBITED AND RESTRICTED WEAPONS, FIREARMS, OR AMMUNITION

Section 101 of the *Criminal Code* gives a peace officer who believes on reasonable grounds that an offence is being committed or has been committed against the *Criminal Code* relating to prohibited and/or restricted weapons, firearms, or ammunition the authority to search without a warrant a person or vehicle or place or premises other than a dwelling-house and seize anything that he or she reasonably believes has been or is being used to commit the offence.

Section 102 of the *Criminal Code* gives a peace officer further authority to seize any restricted weapon in the possession of a person who fails to immediately produce a registration certificate or permission permit for the officer's inspection. This section further authorizes a peace officer to seize a prohibited weapon in the possession of anyone.

Section 103(1) of the *Criminal Code* authorizes the Attorney General or someone on his behalf to apply to a provincial court judge for the issuance of a warrant to search for and seize any firearm or other offensive weapon, or any ammunition or explosive substance where there are reasonable grounds for believing that it is not in the interests of the safety of the person possessing such items or of any other person. Section 103(2) further permits a search and seizure without a warrant for such articles if the peace officer forms the conclusion, based on reasonable grounds, that to apply to a provincial court judge for a search warrant would be impracticable.

GAMING, BETTING, LOTTERIES, AND BAWDY-HOUSES

Section 199(1) of the *Criminal Code* previously permitted a justice of the peace to issue a warrant for the search and seizure of any evidence found in a gaming or betting house, a common bawdy-house, or premises where a lottery was being carried on where the justice received information from an officer that an offence was being committed. In *Re Vella* (1984), 14 C.C.C. (3d) 513 (Ont. H.C.) the court held that section 199(1) was of no force because it was inconsistent with the section 8 Charter protection against unreasonable search and seizure. Section 199 was amended in

1994 (*Criminal Code*, S.C. 1994, c. 44, s. 10) and requires a police officer to satisfy the justice by way of information on oath that such offences are being committed prior to the issuance of a search warrant to search a particular location.

OBSCENE PUBLICATIONS, CRIME COMICS, AND CHILD PORNOGRAPHY

Section 164 of the *Criminal Code* authorizes a judge to issue a warrant to search for and seize any obscene publications, crime comics (section 163(7) provides a definition of "crime comic"), or child pornography (section 163.1(1) provides a definition of "child pornography") that is believed, on reasonable grounds, to be kept for sale or distribution in premises within the jurisdiction of the court.

BLOOD SAMPLES

A person charged with impaired driving arising out of an accident resulting in death or bodily harm to anyone may be required under a warrant or telewarrant issued under section 256 of the *Criminal Code* to allow a qualified medical practitioner to take or to cause to be taken such samples of the person's blood as are necessary to enable a proper analysis to be made of the person's blood alcohol content.

A warrant will only be issued if the justice is satisfied that the accident in question occurred within the preceding two hours. The officer is therefore essentially restricted to the use of a telewarrant given necessary time restrictions. Section 256(4) states that a sample may be taken only when both of the following conditions exist:

1. the medical practitioner is of the opinion that the person is unable to consent because of the accident; and
2. the taking of samples will not endanger the life or health of the person.

If these conditions change during the period between the obtaining of the warrant and the taking of the samples and the person recovers to a degree where consent is possible, the samples may not be taken.

PROCEEDS OF CRIME

Amendments dealing with proceeds of crime were introduced in R.S.C. 1985, c. 42 (4th Supp.), s. 2. The new provisions are contained in Part XII.2 of the *Criminal Code* and were recently amended by S.C. 1997, c. 18, ss. 27-24. This change was proclaimed in 1988 and authorizes the Attorney General to apply to a superior court judge for a warrant to search a building, receptacle, or place for proceeds of crime obtained by a designated drug offence or an enterprise crime, which is defined in the

Criminal Code, section 462.3, and includes offences committed for the benefit of, or in association with, a criminal organization. The application for such a warrant must be in writing and may be *ex parte*. The judge may require any person who has a valid interest in the property to be given notice of the search unless such notice would create conditions that might cause the disappearance of the property or a reduction in its value.

Any such property seized must be preserved and a report must be filed with the clerk of the court carefully identifying the property seized and its location. A copy of this report must also be provided to the person from whom the property was seized and to anyone else whom the judge deems to have a valid interest in it.

PRECIOUS METALS

Section 395 of the *Criminal Code* permits a person who has an interest in a mining claim to lay an information alleging that a precious metal (or rock, mineral, or other substance containing a precious metal) is being unlawfully hidden or held by any person contrary to law. A justice may issue a warrant to search any place or person mentioned in the information.

REGISTERED TIMBER

Section 339(3) of the *Criminal Code* provides a peace officer the right to enter premises without a warrant for the purpose of determining whether or not the timber is being kept with the owner's knowledge or consent if the officer has reasonable grounds to suspect that it is being kept without the owner's knowledge or consent. Unless it is impracticable to do so, the officer should apply for a warrant to search under these circumstances.

POWERS OF SEARCH UNDER OTHER FEDERAL STATUTES

In 1985 section 487 of the *Criminal Code* was amended to extend the use of its search and seizure provisions to other federal statutes. Section 34(2) of the *Interpretation Act* provides:

> All the provisions of the Criminal Code relating to indictable offences apply to indictable offences created by an enactment, and all the provisions of the Code relating to summary conviction offences apply to all other offences created by an enactment, except to the extent that the enactment otherwise provides.

In the case of *R. v. Grant* (1992), 14 C.R. (4th) 260 (B.C.C.A.) it was held that all of the safeguards contained in the federal statute are to be preserved by the police officer. In this case a search warrant issued under the *Criminal Code*, rather than the (now repealed) *Narcotic Control Act*, was found to be invalid as it failed to name the officer executing the warrant, as required under section 12 of the *Narcotic Control Act*.

Section 11(5) of the *Controlled Drugs and Substances Act* provides as follows:

> Where a peace officer who executes a warrant issued under subsection (1) has reasonable grounds to believe that any person found in the place set out in the warrant has on their person any controlled substance, precursor, property or thing set out in the warrant, the peace officer may search that person for the controlled substance, precursor, property or thing and seize it.

POWERS OF SEARCH UNDER PROVINCIAL STATUTES

The powers of the federal Parliament and the provincial legislatures are separated and, therefore, section 487 of the *Criminal Code* does not authorize the issuance of warrants for offences against a provincial statute. As a result, every provincial statute must be examined in order to determine if it contains any power of search and seizure.

The Ontario *Provincial Offences Act* sets out the procedure that must be followed for all provincial offences or by-laws under the authority of an Act of the Ontario Legislature.

Section 158 of the Act confers on justices of the peace powers similar to those powers that pertain to issuing search warrants under the *Criminal Code*. Section 158 authorizes a justice to issue a warrant authorizing a peace officer, etc., to search a building, receptacle, or place for any such thing, once he or she is satisfied by information upon oath.

The search warrant must name the date upon which it expires, not being 15 days after its issue. Warrants must be executed between 6:00 a.m. and 9:00 p.m. (standard time) unless the justice has ordered otherwise in the warrant.

Section 159 prescribes the manner and time limits for the detention of the goods seized and also permits the release of articles for examination and copying.

A search warrant may be issued under section 1(1) of the Act for "an offence under an Act of the Legislature or under a regulation or by-law made under the authority of an Act of the Legislature." In theory, a search warrant is therefore available with regard to any offence under a provincial statute, including offences created by regulation or by municipal by-law.

Individual provincial statutes should be examined to determine if a similar right exists and whether or not it is more extensive than a search warrant under the Ontario *Provincial Offences Act*.

CONCLUSION

This chapter has examined several elements that constitute the law of search and seizure in Canada. It has attempted to show the critical importance of section 8 of the Charter in making determinations about the "reasonableness" of searches and seizures by police officers. We have also been introduced to several leading Canadian cases that establish the foundation of the court's understanding of the principles that pertain to matters of search and seizure.

Additionally, this chapter has reviewed the relevant sections of the *Criminal Code of Canada*, and other statutes, with respect to the police powers of search and seizure. We have included key forms that deal with warrants. It remains for readers to ensure that they stay current in both the case law and statutory amendments dealing with this substantial area of police power.

QUESTIONS FOR CONSIDERATION AND DISCUSSION

1. Consider the aspects of what constitutes a "reasonable" versus an "unreasonable" search. What are the key elements that distinguish the two forms of searches?
2. Discuss the use of telewarrants in Canada. Is it likely that advances in technology will modify, or extend, the use of telewarrants?
3. Review Canadian cases that have addressed the issue of trespass to property in the context of police search and seizure. Are the principles of privacy and the sanctity of the person justified? Will Canadian law be required to consider changes to these principles on the basis of new technology, including advances in medical procedures?
4. After reviewing relevant Canadian criminal cases, discuss the basis for warrantless searches by the police. What circumstances tend to justify such types of searches?

CASES CITED

Cloutier v. Langlois, [1990] 1 S.C.R. 158, 74 C.R. (3d) 316, 53 C.C.C. (3d) 257 (S.C.C.)

Entick v. Carrington (1765), 95 E.R. 807 (K.B.)

Hunter v. Southam Inc. (1984), 41 C.R. (3d) 97 (S.C.C.)

R. v. Borden (1994), 33 C.R. (4th) 147 (S.C.C.)

R. v. Brezack [1949] O.R. 888, 9 C.R. 97 (Ont. C.A.)

R. v. Chapin (1983), 7 C.C.C. (3d) 538 at 541 (Ont. C.A.)

R. v. Church of Scientology (1987), 31 C.C.C. (3d) 447 (Ont. C.A.) leave to appeal to S.C.C. refused (1987), 82 N.R. 393n (S.C.C.)

R. v. Colet (1981), 19 C.R. (3d) 84 (S.C.C.)

R. v. Collins, [1987] 1 S.C.R. 265, 56 C.R. (3d) 193, 33 C.C.C. (3d) 1 (S.C.C.)

R. v. Edwards (1994), 45 C.R. (4th) 307 (S.C.C.)

R. v. Ferguson (1990), 1 C.R. (4th) 53 (Ont. C.A.)

R. v. Godoy (1999), 131 C.C.C. (3d) 129 (S.C.C.)

R. v. Grant (1992), 14 C.R. (4th) 260 (B.C.C.A.)

R. v. Haley (1986), 27 C.C.C. (3d) 454 (Ont. C.A.)

R. v. Hynds (1982), 8 W.C.B. 182 (Alta. Q.B.)

R. v. Mellenthin, [1992] 3 S.C.R. 615

R. v. Moran (1982), 65 C.C.C. (2d) 129 (S.C.C.)

R. v. Posternak (1929), 51 C.C.C. 426 (Alta. C.A.)

R. v. Rao (1984), 46 O.R. (2d) 80, 40 C.R. (3d) 1, 12 C.C.C. (3d) 97 (Ont. C.A.)

R. v. Silveira (1995), 38 C.R. (4th) 330 (S.C.C.)

R. v. Solloway Mills & Co. (1930), 53 C.C.C. 261 (Alta. C.A.)

R. v. Stillman, [1997] 1 S.C.R. 607

R. v. Swarts (1916), 27 C.C.C. 90 (Ont. H.C.)

R. v. Tricker (1995), 96 C.C.C. (3d) 198 (Ont. C.A.)

Radio Canada c. Nouveau-Brunswick (P.G.)(1992), 9 C.R. (4th) 192 (S.C.C.)

Re Laporte (1972), 18 C.R.N.S. 357 (Que. Q.B.)

Re Vella (1984), 14 C.C.C. (3d) 513 (Ont. H.C.)

Re Worrall, [1965] 1 O.R. 527, 44 C.R. 151, [1965] 2 C.C.C. 1 (Ont. C.A.)

Semayne's Case (1604), 77 E.R. 194

Wah Kie v. Cuddy (No. 2) (1914), 23 C.C.C. 383 (Alta. C.A.)

REFERENCES

Canada. Criminal Law Review. Police Powers Project (1986). *Powers and procedures with respect to the investigation of criminal offences and the apprehension of criminal offenders: proposals with commentary*. Ottawa: n.p.

Fontana, James A. (1997). *The law of search and seizure in Canada*. 4th ed. Toronto: Butterworths.

Hutchison, Scott C. (1990–). *Search and seizure law in Canada*. Scarborough, Ont.: Carswell.

Law Reform Commission of Canada (1983). *Police powers: search and seizure in criminal law enforcement*. Ottawa: The Commission.

Law Reform Commission of Canada (1983). *Report on writs of assistance and telewarrants*. Ottawa: The Commission.

Law Reform Commission of Canada (1984). *Report on search and seizure*. Ottawa: The Commission.

Law Reform Commission of Canada (1986). *Report on disposition of seized property: post-seizure procedures*. Ottawa: The Commission.

Law Reform Commission of Canada (1991). *Recodifying criminal procedure. Vol. 1 police powers: title 1, search and related matters: highlights of recommendations*. Ottawa: The Commission.

Manning, Morris (1983). *Rights, freedoms and the courts: a practical analysis of the Constitution Act, 1982*. Toronto: Emond-Montgomery.

McCalla, Winston (1984). *Search and seizure in Canada*. Aurora, Ont.: Canada Law Book.

Paiken, Lee (1980). *The issuance of search warrants: a manual prepared for the Law Reform Commission of Canada*. Ottawa: The Commission.

Salhany, Roger E. (1997). *Police manual of arrest, seizure & interrogation*. 7th ed. Scarborough, Ont.: Carswell.

Stenning, Philip C. and Clifford D. Shearing (1979). *Search and seizure of private security personnel: a study paper prepared for the Law Reform Commission of Canada*. Hull, Quebec: Minister of Supply and Services Canada.

RELATED ACTIVITIES

- Review media coverage of Canadian cases dealing with the police powers of search and seizure.
- Complete applications for, and actual, search warrants for specific criminal circumstances using the forms authorized under the *Criminal Code*.
- Research current releases of *Canadian Criminal Cases* and/or *Criminal Reports* for topical cases dealing with:
  - Appeals under section 8 of the Charter;
  - Telewarrants; and
  - Warrantless searches.
- Acquire training manuals, police policies, guidelines, and/or procedures dealing with search and seizure from different police departments. They could be within your own jurisdiction, or, from other jurisdictions.

WEBLINKS

 http://www.rcmp-grc.gc.ca/html/te-crime3xx.htm Go to this site to read an interesting case study on search warrants and the internet.

 http://www.sgc.gc.ca/EPub/Pol/eSurveillance99/eSuveillance99.htm This site details electronic Surveillance Guidelines for Agents and Peace Officers designated by the Solicitor General of Canada guidelines on electronic surveillance used to intercept private communication.

 http://www.sgc.gc.ca/epub/pol/e199611/e199611.htm This paper explores the issues surrounding the DNA collection and warrants.

CHAPTER FOUR

INTERIM RELEASE OF
ARRESTED PERSONS

LEARNING OBJECTIVES

1. State the principle supporting interim, or pre-trial, release of accused persons in Canadian law.
2. Describe the application of statutory authority for the interim release of arrested persons by a peace officer.
3. Describe the application of statutory authority for the interim release of arrested persons by an officer in charge.
4. Describe the application of statutory authority for judicial interim release of arrested persons.

INTRODUCTION

This chapter will deal with the subject of pre-trial, or interim, release as it pertains to a "peace officer," an "officer in charge," and a judge or justice in Canadian law.

The origin of the concept of bail will be addressed as a precursor to interim release. Also, because the hierarchical nature of the scheme developed to guide pre-trial release is relatively complex, this chapter will present substantial portions from the *Criminal Code* (including appropriate Forms) that set out the relevant provisions in this area of law.

FROM BAIL TO PRE-TRIAL, OR INTERIM, RELEASE

It has been noted that the concept of bail, which has been replaced in current Canadian legal language by the term "pre-trial release," predates recorded English law (Delisle and Stuart, 1996; Trotter, 1999). As Delisle and Stuart indicate:

> The notion of bail pending trial antedates recorded English law and its original purpose is therefore not altogether certain. It may have stemmed from the medieval sheriff's desire to avoid

the costly and troublesome burden of personal responsibility for those in his charge. Whether motivated by a concern for their prisoner's well-being or as an economy measure, sheriffs commonly released prisoners either on their own recognizances, with or without requiring the posting of some sort of bond, or on the promise of a third party to assume personal responsibility for the accused's appearance at trial. From ad hoc arrangements by the sheriff we move by the Statute of Westminster of 1275 to a system wherein conditions for pre-trial release are specified and the power of release [is] transferred from the sheriff to justice of the peace. To ensure that the accused would reappear on the date set for trial, a third party, or surety, had to assume a personal responsibility for the accused, on penalty of forfeiture of his own property. These sureties, usually local landowners, were given the powers of a jailer to prevent the accused's flight and this system was eminently reasonable in an immobile land-oriented society. (p. 246)

English law made adjustments to account for the reality of the inordinately long delays endured by persons in custody awaiting trial. Frequently an accused person would be held indefinitely without any degree of certainty as to when his or her case would be heard before a judge. In earlier times, the justices would travel from town to town, holding circuit courts to deal with numerous legal matters. This arrangement, and the unfairness and cruelty to which it was prone, required a new system in order to approach the demands of justice. As Trotter (1999) indicates from his research:

Many inmates died while awaiting (sometimes for a number of years) the arrival of itinerant justices who would preside over their trials . . . During this period of time, arrested persons were the responsibility of local sheriffs, who were the King's representatives. As the sheriffs were liable to substantial fines each time a prisoner escaped, it is not surprising that they sought to divest themselves of this responsibility.

From this untenable situation, a system emerged whereby prisoners would be released to their family or friends, who would guarantee to produce the accused person in court to face the allegations. The guarantee usually entailed some type of financial obligation. As such, local landowners and knights were the most desirable persons to assume this role. (p. 3)

However, the system that was devised was subject to its own abuses. Because of the substantial discretion given to the sheriffs, there was uncertainty over who was, and who was not, to be released. Furthermore, and perhaps not surprisingly, there were often charges that the sheriffs were open to bribery and influence in their individual approach to pre-trial release (Trotter, 1999, p. 3).

In Canada, we have inherited from English law a substantial amount of criminal procedure as it pertained to bail, or interim judicial release. It was originally based on the essential need to ensure that the accused person would appear in court. Gradually, additional considerations were taken into account in making determinations on the granting of bail (Trotter, 1999). These considerations included:

- the seriousness of the offence;
- the severity of any penalty upon a finding of guilt;
- the strength of the evidence against the accused; as well as,

- the character of the accused and his or her standing within the community.

One Canadian case, out of British Columbia, that dealt with the issue of the accused's likely attendance at court is *R. v. Gottfriedson* (1906), 10 C.C.C. 239 (B.C. Co. Ct.).

Before Canadian Confederation, in 1867, the law of bail mirrored English practice. Accordingly, bail was considered a right with respect to misdemeanours and discretionary for those offences considered as felonies (Trotter, 1999). However, with the enactment of Canada's first criminal legislation package, in 1869, bail was made discretionary for all offences (Mewett, 1967; Parker, 1981).

However, the law of bail in Canada was dramatically altered with the gradual recognition of other factors, beyond securing an accused's attendance at court, as valid grounds for detaining a person. This recognition has repercussions today and has added complexity with the appearance of the *Canadian Charter of Rights and Freedoms*. The genesis of this change has been identified by Trotter (1999):

> The impetus for this Canadian development was the decision of the English Court of Appeal in R. v. Phillips (1947). In this case, the Court considered an appeal against a sentence of four years in respect of a number of burglaries. The Court took into consideration a number of offences that Phillips had committed while on bail. The Court was also apprised of Phillip[s'] serious record of burglary and property related crimes. (p. 8)

The court, in dismissing the appeal, felt that Mr. Phillips (*R. v. Phillips* (1947), 32 Cr. App. Rep. 47 (C.C.A.)) should not be released on bail. Bail is a matter of discretion in cases of felony.

It was agreed that certain crimes are not likely to be repeated pending trial. In those cases there should be no objection to bail. However, in other cases, that likelihood does exist. It was considered that housebreaking, the crime for which Mr. Phillips was charged, was indeed such a crime. Such crimes will very probably be repeated if a prisoner is released on bail, especially in the case of someone like Phillips, who had a record for housebreaking. It was seen as an offence that can be committed with a considerable measure of safety to the person committing that crime. The court felt that to turn Phillips loose on society was "a very inadvisable step."

Other issues that should be taken into account in the decision to grant or deny bail include:

- the nature of the accusation;
- the nature of the evidence in support of the accusation; and
- the severity of the punishment which conviction will entail.

As a result, in Canada, legislation was introduced that modified the approach to bail and created a system of pre-trial release that has been with us, with modifications,

ever since. The *Bail Reform Act* and the *Criminal Law Amendment Act, 1968-69,*
S.C. 1968-69, c. 38, s. 31 changed the approach to release after arrest (Trotter, 1999).
These pieces of legislation introduced "a complex hierarchical scheme of police pow-
ers to compel an accused's appearance instead of or after arrest" (Trotter, 1999, p. 4).
This scheme provides some protection for peace officers against criminal liability
(Trotter, 1999):

> Although this hierarchical scheme imposes a duty either not to arrest or to release after ar-
> rest, a failure to adhere to it does not jeopardize the authority of the peace officer. Where the
> peace officer or officer in charge fails to release in accordance with these duties, he or she is
> nonetheless deemed to be acting in the execution of duty so as to avoid criminal, if not neces-
> sarily civil, liability. (p. 6)

Essentially, this scheme combines a power to arrest with corresponding duties
not to arrest or to release after arrest. These duties are balanced within the context of
two key factors:

1. the status of the peace officer either as a peace officer or an officer in charge;
   and
2. the type of crime alleged to have been committed.

On the face of it, the scheme appears to confer on more senior officers an obligation for dealing with the more serious crimes. However, the scheme is not so simple. The *Criminal Code* introduces a degree of complexity into the scheme by allowing
for an apparently broad discretionary power to release that clouds the distinction between a "peace officer" and an "officer in charge," as well as, the distinction between
less serious and more serious crimes.

Trotter (1999) has addressed the changes that clarified the powers of the police in
Canada to release an accused person:

> Although the powers of the police to take a person into custody have always been reasonably clear,
> the issue of whether a police officer could release an arrested person without first taking that
> person before a justice was uncertain for some time. This was largely due to the fact that the
> Criminal Code was silent on this issue. However, a practice developed in England whereby
> police officers could release prisoners on a form of bail. The existence of this power in Canada
> was disputed by some commentators. The debate ended in 1969 when Parliament specifically
> conferred on police officers that power to release an arrested accused with the intention of
> compelling his or her appearance by summons. Hence, the emergence of police bail in Canada.
> (p. 37)

A judicial authority to compel the appearance of an accused person generally arose
after a person (typically a peace officer) laid an information before a justice. Historically,
once that information was laid, and a case established for issuing a process at an *ex parte*
hearing (i.e., where only one side is heard from), a justice could issue a summons, or

a warrant, to compel appearance. With the introduction of the *Bail Reform Act*, however, this approach was changed. This Act created another avenue for the laying of an information whereby a peace officer, or officer in charge, could issue an appearance notice, a promise to appear, or a recognizance. Also, the Act introduced a system imposing specific rules on justices with respect to the manner of compelling an accused's appearance (i.e., by summons or warrant), release, or detention (Law Reform Commission of Canada, 1988).

THE PURPOSE OF POLICE CUSTODY PRIOR TO FIRST APPEARANCE

The need to have accused persons taken into custody upon arrest, and prior to first appearance before a judge, is important to understand in the continuum of the criminal justice system. Regardless of the charge, irrespective of the status of the accused, and notwithstanding our general allegiance to the principle of presumption of innocence, police custody subsequent to arrest has strong justification. The Law Reform Commission of Canada (1988) offered the following explanation for this important police practice:

> Police custody prior to first appearance serves two purposes. First, it is valuable for police investigation. During this period, police regularly question and search the accused or conduct a variety of investigative tests. Second, it protects the public from persons, who it is believed, would jeopardize the public interest. It follows logically that police custody must satisfy the need to maintain security and order in the place of custody. (p. 20)

Credit: Mike Weaver, Media Relations, Kingston Police

An accused person being held in custody.

Scollin (1977) also summarizes the need for police to take an accused person into temporary custody for specific organizational purposes that serve the overall goals of the criminal justice system:

> For proper investigative purposes such as identification, prevention of flight and search and seizure of evidence to prevent it being destroyed, arrests without warrant will continue to be made. After the immediate purposes of law enforcement have been served, however, it will frequently no longer be necessary to hold the alleged offender in custody until a justice is available. (pp. 3–4)

FAIRNESS, THE CHARTER, AND PUBLIC SAFETY

The reader will note that Scollin, above, makes an important qualification when speaking about the release of an accused person once the investigative purposes of law enforcement have been met. Beyond compelling the accused's appearance at trial, there are other significant considerations that must be taken into account prior to the release of an alleged offender. However, these considerations must operate within the framework of our understanding of fairness and due process. In their review of the existing process of compelling appearance, interim release, and pre-trial detention, the Law Reform Commission of Canada (1988) makes the following observation:

> Fairness requires that detention should be used as a last resort. As the Canadian Committee on Corrections pointed out, unjustified detention shows disregard for human rights. Lack of segregation from those already convicted of crime, stringent security measures, disruption of family and other social contacts—all these can harm the person detained. (p. 27)

The impact of this segregation and separation on the accused person must be considered in the context of the decision to release. It has been established through research that persons held in custody pending trial are often dealt with more harshly by the criminal justice system than are those granted bail. For example, Professor Martin Friedland, from the University of Toronto Law School, conducted an empirical study of bail procedures in the Toronto Magistrates' Court. Trotter (1999) summarizes the results of Friedland's important study:

> This classic work drew attention to many problems with the administration of the bail provisions in the Criminal Code. Among his criticisms of the system, Friedland observed that the existing procedures for compelling the attendance of the accused in court resulted in the detention of many individuals whose attendance could have been secured by less intrusive means. Friedland also raised serious questions about the fairness of requiring security in advance as a condition precedent to release. Most importantly, perhaps, was Friedland's identification of the "disturbing relationship" between pre-trial detention and the outcome of trial. In particular, this study found that a detained accused was more likely than his or her bailed counterpart to be convicted of the offence charged. Furthermore, a detained person was more likely to receive a custodial sentence. Simply put, accused persons who were denied bail received more jail, more often. (p. 11)

The Law Reform Commission of Canada's (1988) review of matters pertaining to compelling appearance, interim release, and pre-trial detention acknowledged the important relationship between the decision-making surrounding these practices and the *Charter of Rights and Freedoms*:

> Unlike correctional law, the law of criminal procedure has a narrower ambit. It is designed to ensure that a person whose liberty is put at risk can have a full and fair trial of the charge made against him or her. In this context, obviously Charter and other guarantees are important. For example, the Charter provides a right to counsel without delay, the right to be informed properly of the reasons for detention or arrest, the right to habeas corpus and the right not to be subjected to cruel and unusual punishment. (p. 21)

Again, as the Law Reform Commission of Canada (1988) notes in this context:

> When a crime is committed, the state must be able to invoke a process whereby the actions of the accused can be judged in a fair and impartial manner. For the process to work, the criminal justice system must have the power to compel the attendance of the accused or of a witness whose evidence is needed. (p. 27)

The question then arises: what are the best means available to ensure that the accused person will attend court? This is followed by the additional important question: when should an accused person be released prior to trial? The *Criminal Code* provisions under Part XVI constitute the answers to those questions.

The Law Reform Commission of Canada's study also reveals the importance of taking a position that is likely to be consistent with the overriding principles of the Charter. This is motivated by considerations of fairness and practicality:

> Clearly the law of interim release and detention should, whenever possible, avoid . . . ambiguity. For purposes of policy formulation, the best means to achieve this is to interpret the Charter in a liberal manner. This approach achieves consistency with the spirit and intent of the Charter by affirming its legal guarantees rather than avoiding them by resort to limitations of questionable legality. Moreover, it is also a practical approach since a statutory scheme infused with a broad interpretation of these guarantees should not be open to successful challenges on the basis that they have been contravened. It therefore avoids the cost to the criminal justice system of lengthy Charter challenges. (p. 37)

Trotter considers the willingness of the Quebec Court of Appeal, in the face of initial judicial reluctance, to invoke the Charter, and particularly the presumption of innocence provisions in making determinations with respect to bail:

> Despite this early, cautious approach of the courts to the application of the Charter to the law of bail, the Quebec Court of Appeal has recently tackled bail issues under the Charter with vigour. In the context of interpreting the meaning of the "public interest" in s. 515(10)(b), the Court has tempered previous judicial pronouncements on this issue by meaningfully injecting the presumption of innocence into the analysis. In this sense, the Charter does not merely elevate existing constructs to the level of constitutional entitlements; it also transforms them in the process. (pp. 16–17)

This issue was dealt with in the case of *R. v. Lamothe* (1990), 58 C.C.C. (3d) 530 (Que. C.A.).

An interesting focus of discussion has arisen with regard to the ambiguity concealed in the term "public interest." Several cases have examined this concept in some detail and have attempted to clarify an appropriate definition, as well as, to distinguish it from a phrase such as "protection or safety of the public." The following cases are relevant in the context of this discussion:

- *Re Powers and the Queen* (1972), 9 C.C.C. (2d) 533 (Ont. H.C.);
- *R. v. Demyen* (1975), 26 C.C.C. (2d) 324 (Sask. C.A.); and
- *R. v. Kingswatsiak* (1976), 31 C.C.C. (2d) 213 (N.W.T.C.A.).

Here the concept of "public interest" was defined to include the notion of the "public image" of the *Criminal Code* and its importance in sustaining a high degree of public confidence in the administration of justice in Canada.

In his review of the law of bail in Canada, Trotter (1999) offers the following guidance:

> The "public interest" is a broad and nebulous concept. As a criminal law standard for adjudicating upon the liberty of the accused, its imprecision is unsatisfactory in some respects. However, it does represent a theme that is important to the criminal justice system—a consideration of the impact of certain decisions on the perception of the integrity of the system. (p. 91)

One interesting aspect of the public interest relates to the nature of the community in which the alleged crime has taken place. It is proposed that there may be a distinction between a small community and a large, urban area with respect to any determinations that are made about permitting the pre-trial release of an accused person. Again, Trotter (1999) is helpful in his articulation of this consideration:

> The nub of this concern is engaged when a particularly disturbing crime occurs in a small or isolated community. It may be said that a horrendous crime committed in a small community may have more of an impact than one committed in a large, urban centre. It is the community into which the accused is released that will have to live with the accused. In a small community, many members may know the accused and the victims personally. Feelings could run high. In a large municipality, the released accused will likely fade into the community with anonymity, causing little grief to most people. Should the public interest be approached differently in these two situations? (p. 101)

PRINCIPLES OF ACCOUNTABILITY, EFFICIENCY, AND CLARITY

Beyond the issues that relate to fairness to the accused, the Charter, and public safety, there are further concerns with respect to the principles under which the interim release of accused persons should operate. Several authors have examined the following topics (Law Reform Commission of Canada, 1988; Scollin, 1977; Trotter, 1999):

- Accountability;
- Efficiency; and
- Clarity.

In the category of accountability, it is essential that those who wield the power to grant, or deny, interim release to an accused person should be clearly held to account for the decisions they make in this regard. Therefore, peace officers, officers in charge (for an examination of the meaning of "officer in charge" the reader is referred to the case of *R. v. Gendron* (1985), 22 C.C.C. (3d) 312 (Ont. C.A.)), and justices must adhere to significant standards of accountability that will warrant the trust of those who are subject to these laws, as well as, the public at large. This point is expressed clearly in the study conducted by the Law Reform Commission of Canada (1988):

> The principle of accountability requires that those exercising procedural power of authority should be accountable for its use so as to inhibit the possibility of abuse. While the law must permit state officials to exercise discretion, it should consciously avoid providing opportunities for the exercise of arbitrary power. Thus, accountability ensures conformity to the law by providing remedies for those affected by the interim release; this principle mandates that where the law imposes duties upon those enforcing interim-release procedures, a failure to adhere to them should give rise to a remedy. (p. 28)

By making the rules regarding interim release detailed enough to ensure appropriate levels of accountability, there is also a goal that speaks to efficiency. Through an approach that places specific tasks upon identified agents within the criminal justice system, there is an intention to achieve an efficient process. In addition, there is an interest in making the approach taken clear. Again, the Law Reform Commission (1988) is useful in addressing the matter of clarity:

> Coupled with efficiency is the principle of clarity. This is a necessary underpinning to the concept of the rule of law. Ideally, the law should guide others so that one knows what is or is not permissible. Thus, any scheme of interim release, just like any other rule of criminal procedure, must be both comprehensive and understandable. Comprehensive, because otherwise gaps in the law would give rise to confusion about what the law is. Understandable, so that persons may use the law to guide their actions. (p. 29)

RECOGNIZANCE

Trotter (1999) provides the following explanation for the term "recognizance," which appears frequently in the sections of the Code dealing with interim release:

> The term "recognizance" is a technical term of art which is not defined in the Criminal Code, even though Form 32 is titled "Recognizance". Despite the lack of statutory assistance in ascribing a definition to this term, its meaning is deeply rooted in the common law. Generally, a recog-

nizance is the formal record of an acknowledgement of indebtedness to the Crown which is defeasible upon the fulfilment of certain conditions, the primary one being attendance in court for trial. The actual document that is executed, the recognizance, does not create the obligation, but merely records it. (p. 161)

By entering into a recognizance, the person is verifying that he or she is liable to pay a set amount of money to Her Majesty the Queen upon failing to comply with the specific conditions contained in that document (e.g., attending at his or her trial). The accused must acknowledge this obligation by signing Form 11, reproduced below in Figure 4.1.

FIGURE 4.1

FORM 11 – RECOGNIZANCE ENTERED INTO BEFORE AN OFFICER IN CHARGE OR OTHER PEACE OFFICER
(Section 493)

Canada,
Province of ......,
(territorial division).

I, A.B., of ....., *(occupation)*, understand that it is alleged that I have committed *(set out substance of offence)*.

In order that I may be released from custody, I hereby acknowledge that I owe $ *(not exceeding $500)* to Her Majesty the Queen to be levied on my real and personal property if I fail to attend court as hereinafter required. *(or, for a person not ordinarily resident in the province in which the person is in custody or within two hundred kilometres of the place in which the person is in custody)*

In order that I may be released from custody, I hereby acknowledge that I owe $ *(not exceeding $500)* to Her Majesty the Queen and deposit herewith *(money or other valuable security not exceeding the amount or value of $500)* to be forfeited if I fail to attend court as hereinafter required.

1. I acknowledge that I am required to attend court on ...... day, the ......day of .... A.D. ...., at .... o'clock in the ....noon, in courtroom No. ...., at the court, in order to be dealt with according to law.
2. I acknowledge that I am also required to appear on .... day, the .... day of .... A.D. ...., at .... o'clock in the .... noon, at *(police station)*, *(address)*, for the purposes of the *Identification of Criminals Act*. (Ignore if not filled in).

I understand that failure without lawful excuse to attend court in accordance with this recognizance to appear is an offence under subsection 145(5) of the *Criminal Code*.

Section 145(5) and (6) of the *Criminal Code* state as follows:

"(5) Every person who is named in an appearance notice or promise to appear, or in a recognizance entered into before an officer in charge or another peace officer, that has been confirmed by a justice under section 508 and who fails, without lawful excuse, the proof of which lies on the person, to appear at a time and place stated therein, if any, for the purposes of the Identification of Criminals Act or to attend court in accordance therewith, is guilty of

(a)   an indictable offence and liable to imprisonment for a term not exceeding two years; or

(b)   an offence punishable on summary conviction.

(6) For the purposes of subsection (5), it is not a lawful excuse that an appearance notice, promise to appear or recognizance states defectively the substance of the alleged offence."

Section 502 of the *Criminal Code* states as follows:

"

502. Where an accused who is required by an appearance notice or promise to appear or by a recognizance entered into before an officer in charge or another peace officer to appear at a time and place stated therein for the purposes of the Identification of Criminals Act does not appear at that time and place, a justice may, where the appearance notice, promise to appear or recognizance has been confirmed by a justice under section 508, issue a warrant for the arrest of the accused for the offence with which he is charged."

Dated this …….. day of ……… A.D. ………..., at ……………

....
(Signature of accused)
1997, c. 18, s. 115

The accused is then brought before a judge or justice where the terms and conditions of the recognizance are recorded in Form 32, reproduced in Figure 4.2 below.

FIGURE 4.2

FORM 32 – RECOGNIZANCE
(Sections 493, 550, 679, 706, 707, 810.1 and 817)

Canada,

Province of ....,
(*territorial division*)

Be it remembered that on this day the persons named in the following schedule personally came before me and severally acknowledged themselves to owe to Her Majesty the Queen the several amounts set opposite their respective names, namely,

Name	Address	Occupation	Amount
A.B.			
C.D.			
E.F.			

to be made and levied of their several goods and chattels, lands and tenements, respectively, to the use of Her Majesty the Queen, if the said A.B. fails in any of the conditions herein written.

Taken and acknowledged before me on the .... day of .... A.D. .... , at
....

....
Judge, Clerk of the Court,
Provincial Court Judge *or* Justice

1. Whereas the said ...., hereinafter called the accused, has been charged that (*set out the offence in respect of which the accused has been charged*);

Now, therefore, the condition of this recognizance is that if the accused attends court on .... Day, the .... day of .... A.D. .... , at ....
o'clock in the .... noon and attends thereafter as required by the court in order to be dealt with according to law (*or, where date and place of appearance before the court are not known at the time recognizance is entered into if the accused attends at the time and place fixed by the court and attends thereafter as required by the court in order to be dealt with according to the law*) [515, 520, 521, 522, 523, 524, 525, 680];
And further, if the accused (*insert in Schedule of Conditions any additional con-*

ditions that are directed), the said recognizance is void, otherwise it stands in full force and effect.

2. Whereas the said ...., hereinafter called the appellant, is an appellant against his conviction (*or* against his sentence) in respect of the following charge (*set out the offence for which the appellant was convicted*) [679, 680];

Now, therefore, the condition of this recognizance is that if the appellant attends as required by the court in order to be dealt with according to law;

And further, if the appellant (*insert in Schedule of Conditions any additional conditions that are directed*), the said recognizance is void, otherwise it stands in full force and effect.

3. Whereas the said ...., hereinafter called the appellant, is an appellant against his conviction (*or* against his sentence or against an order *or* by way of stated case) in respect of the following matter (*set out offence, subject-matter of order or question of law*) [816, 831, 832, 834];

Now, therefore, the condition of this recognizance is that if the appellant appears personally at the sittings of the appeal court at which the appeal is to be heard;

And further, if the appellant (*insert in Schedule of Conditions any additional conditions that are directed*), the said recognizance is void, otherwise it stands in full force and effect.

4. Whereas the said ...., hereinafter called the appellant, is an appellant against an order of dismissal (*or* against sentence) in respect of the following charge (*set out the name of the accused and the offence, subject-matter of order or questions of law*) [817, 831, 832, 834];

Now, therefore, the condition of this recognizance is that if the appellant appears personally or by counsel at the sittings of the appeal court at which the appeal is to be heard the said recognizance is void, otherwise it stands in full force and effect.

5. Whereas the said ...., hereinafter called the accused, was ordered to stand trial on a charge that (*set out the offence in respect of which the accused has been charged*);

And whereas A.B. appeared as a witness on the preliminary inquiry into the said charge [550, 706, 707];

Now, therefore, the condition of this recognizance is that if the said A.B. appears at the time and place fixed for the trial of the accused to give evidence on the indictment that is found against the accused, the said recognizance is void, otherwise it stands in full force and effect.

6. The condition of the above written recognizance is that if A.B. keeps the peace and is of good behaviour for the term of .... commencing on ...., the said recognizance is void, otherwise it stands in full force and effect [810 and 810.1].

7. Whereas a warrant was issued under section 462.32 or a restraint order was made under subsection 462.33(3) of the *Criminal Code* in relation to any property (*set out a description of the property and its location*);

Now, therefore, the condition of this recognizance is that A.B. shall not do or cause anything to be done that would result, directly or indirectly, in the disappearance, dissipation or reduction in value of the property or otherwise affect the property so that all or a part thereof could not be subject to an order of forfeiture under section 462.37 or 462.38 of the *Criminal Code* or any other provision of the *Criminal Code* or any other Act of Parliament [462.34].

Schedule of Conditions

(a) reports at (*state times*) to (*name of peace officer or other person designated*),

(b) remains within (*designated territorial jurisdiction*),

(c) notifies (*name peace officer or other person designated*) of any change in his address, employment or occupation,

(d) abstains from communicating, directly or indirectly, with (*identification of victim, witness or other person*) except in accordance with the following conditions: (*as the justice or judge specifies*);

(e) deposits his passport (*as the justice or judge directs*), and

(f) (*any other reasonable conditions*).

The case of *Bietel v. Ouseley* (1921), 35 C.C.C. 386 (Sask. C.A.) addresses the question of a recognizance from a Canadian perspective. A justice or judge may also require that the accused person provide sureties before being released on a recognizance.

Special care must be taken by police officers when dealing with young offenders.

Those persons identified as sureties will forfeit their own funds, to the specific amount agreed to, should the accused fail to comply with the elements of the recognizance.

The justice or judge may require that the accused obtain sureties before being released on recognizance. If there are sureties, they will forfeit their own money in the amount agreed upon should the accused fail to comply with the recognizance to the degree that they are at fault. The following two Canadian cases address the issue of sureties:

- *R. v. Andrews* (1975) 34 C.R.N.S. 344 (Nfld. S.C.); and
- *R. v. Sandhu* (1984), 38 C.R. (3d) 56 (Que. S.C.).

GROUNDS AND CONDITIONS FOR RELEASE

Trotter (1999) has summarized the various grounds and conditions for release as found in the *Criminal Code*:

PRIMARY GROUNDS FOR RELEASE

1. the nature of the offence and the potential penalty;
2. the strength of the evidence against the accused;
3. the ties the accused has to the community;

4. the character of the accused person;
5. the accused's record of compliance with court orders on previous occasions; and
6. the accused's behaviour prior to apprehension (e.g., no evidence of flight during the arrest).

CONDITIONS FOR RELEASE

The conditions for release may include monetary and non-monetary conditions. The non-monetary conditions may include:

1. Enumerated conditions—
   • report to peace officer
   • remain within a territorial jurisdiction
   • notify of changes in address and employment
   • abstain from communication with anyone identified by the court
   • deposit passport
   • comply with firearms prohibition; and
2. Non-enumerated conditions—
   • geographical restrictions
   • curfews
   • control of drug or alcohol consumption
   • driving prohibitions
   • medical treatment
   • possessing bail papers
   • picketing or demonstration prohibitions.

The following text reproduces excerpts from the *Criminal Code of Canada* that are relevant to the concepts of judicial and interim release. It is important to emphasize that these sections are current as of the publication of this textbook and may be modified through subsequent amendments.

RELEVANT SECTIONS FROM THE *CRIMINAL CODE OF CANADA*

PART XVI—COMPELLING APPEARANCE OF AN ACCUSED BEFORE A JUSTICE AND INTERIM RELEASE

493. **Definitions**—In this Part

"accused" includes

(a) a person to whom a peace officer has issued an appearance notice under section 496, and

(b) a person arrested for a criminal offence;

"appearance notice" means a notice in Form 9 issued by a peace officer;

"judge" means

(a) in the Province of Ontario, a judge of the superior court of criminal jurisdiction of the Province,
(b) in the Province of Quebec, a judge of the superior court of criminal jurisdiction of the province or three judges of the Court of Quebec,
(c) [Repealed 1992, c.51, s. 37]
(d) in the Provinces of Nova Scotia, New Brunswick, Manitoba, British Columbia, Prince Edward Island, Saskatchewan, Alberta and Newfoundland, a judge of the superior court of criminal jurisdiction of the Province,
(e) in the Yukon Territory and the Northwest Territories, a judge of the Supreme Court of the territory, and
(f) in Nunavut, a judge of the Nunavut Court of Justice;

"officer in charge" means the officer for the time being in command of the police force responsible for the lock-up or other place to which an accused is taken after arrest or a peace officer designated by him for the purposes of this Part who is in charge of that place at the time an accused is taken to that place to be detained in custody;

"promise to appear" means a promise in Form 10

"recognizance," when used in relation to a recognizance entered into before an officer in charge, or other peace officer, means a recognizance in Form 11, and when used in relation to a recognizance entered into before a justice or judge means a recognizance in Form 32;

"summons" means a summons in Form 6 issued by a justice or a judge;

"undertaking" means an undertaking in Form 11.1 or 12;

"warrant," when used in relation to a warrant for the arrest of a person, means a warrant in Form 7 and, when used in relation to a warrant for the committal of a person means a warrant in Form 8.

RELEASE FROM CUSTODY BY PEACE OFFICER

497. (1) Release from custody by peace officer—Subject to subsection (1.1), if a peace officer arrests a person without warrant for an offence described in paragraph 496(a), (b) or (c), the peace officer shall, as soon as practicable,

(a) release the person from custody with the intention of compelling their appearance by way of summons; or

(b) issue an appearance notice to the person and then release them.

Exception—A peace officer shall not release a person under subsection (1) if the peace officer believes, on reasonable grounds,

(a) that it is necessary in the public interest that the person be detained in custody or that the matter of their release from custody be dealt with under another provision of this Part, having regard to all the circumstances including the need to
   (i) establish the identity of the person,
   (ii) secure or preserve evidence of or relating to the offence,
   (iii) prevent the continuation or repetition of the offence or the commission of another offence, or
   (iv) ensure the safety and security of any victim of or witness to the offence; or
(b) that if the person is released from custody, the person will fail to attend court in order to be dealt with according to law.

RELEASE FROM CUSTODY BY OFFICER IN CHARGE

498. (1) Release from custody by officer in charge—Subject to subsection (1.1), if a person who has been arrested without warrant by a peace officer is taken into custody, or if a person who has been arrested without warrant and delivered to a peace officer under subsection 494(3) or placed in the custody of a peace officer under subsection 163.5 of the Customs Act is detained in custody under subsection 503(1) for an offence described in paragraph 496(a), (b) or (c), or any other offence that is punishable by imprisonment for five years or less, and has not been taken before a justice or released from custody under any other provision of this Part, the officer in charge or another peace office shall, as soon as practicable,

(a) release the person with the intention of compelling their appearance by way of summons;
(b) release the person on their giving a promise to appear;
(c) release the person on the person's entering into a recognizance before the officer in charge or another peace officer without sureties in an amount not exceeding $500 that the officer directs, but without deposit of money or other valuable security; or
(d) if the person is not ordinarily resident in the province in which the person is in custody or does not ordinarily reside within 200 kilometres of the place in which the person is in custody, release the person on the person's entering into a recognizance before the officer in charge or another peace officer without sureties in an amount not exceeding $500 that the officer directs and, if

the officer so directs, on depositing with the officer a sum of money or other valuable security not exceeding in amount or value $500, that the officer directs.

(1.1) **Exception**—The officer in charge or the peace officer shall not release a person under subsection (1) if the officer in charge or peace officer believes, on reasonable grounds,

(a) that it is necessary in the public interest that the person be detained in custody or that the matter of their release from custody be dealt with under another provision of this Part, having regard to all the circumstances including the need to
   (i) establish the identity of the person,
   (ii) secure or preserve evidence of or relating to the offence,
   (iii) prevent the continuation or repetition of the offence or the commission of another offence, or
   (iv) ensure the safety and security of any victim of or witness to the offence; or
(b) that, if the person is released from custody, the person will fail to attend court in order to be dealt with according to law.

499. (1) Release from custody by officer in charge where arrest made with warrant—Where a person who has been arrested with a warrant by a peace officer is taken into custody for an offence other than one mentioned in section 522, the officer in charge may, if the warrant has been endorsed by a justice under subsection 507(6),

(a) release the person on the person's giving a promise to appear;
(b) release the person on the person's entering into a recognizance before the officer in charge without sureties in the amount not exceeding five hundred dollars that the officer in charge directs, but without deposit of money or other valuable security; or
(c) if the person is not ordinarily resident in the province in which the person is in custody or does not ordinarily reside within two hundred kilometres of the place in which the person is in custody, release the person on the person's entering into a recognizance before the officer in charge without sureties in the amount not exceeding five hundred dollars that the officer in charge directs, and if the officer in charge so directs, on depositing with the officer in charge such sum of money or other valuable security not exceeding in amount or value five hundred dollars, as the officer in charge directs.

(2) Additional conditions—In addition to the conditions for release set out in paragraphs (1)(a), (b) and (c), the officer in charge may also require the person to enter into an undertaking in Form 11.1 in which the person, in order to be released, undertakes to do one or more of the following things:

(a) to remain within the territorial jurisdiction specified in the undertaking;
(b) to notify a peace officer or another person mentioned in the undertaking of any change in his or her address, employment or occupation;
(c) to abstain from communicating directly or indirectly, with any victim, witness or other person identified in the undertaking, or from going to a place specified in the undertaking, except in accordance with the conditions specified in the undertaking;
(d) to deposit the person's passport with the peace officer or other person mentioned in the undertaking;
(e) to abstain from possessing a firearm and to surrender any firearm in the possession of the person and any authorization, licence or registration certificate or other document enabling that person to acquire or possess a firearm;
(f) to report at the times specified in the undertaking to a peace officer or other person designated in the undertaking;
(g) to abstain from
    (i)   the consumption of alcohol or other intoxicating substances,
    (ii) the consumption of drugs except in accordance with a medical prescription; and
(h) to comply with any other conditions specified in the undertaking that the officer in charge considers necessary to ensure the safety and security of any victim of or witness to the offence.

503. (1) Taking before justice—A peace officer who arrests a person with or without warrant or to whom a person is delivered under subsection 494(3) or into whose custody a person is placed under subsection 163.5(3) of the Customs Act shall cause the person to be detained in custody and, in accordance with the following provisions, to be taken before a justice to be dealt with according to law:

(a) where a justice is available within a period of twenty-four hours after the person has been arrested by or delivered to the peace officer, the person shall be taken before a justice without unreasonable delay and in any event within that period, and
(b) where a justice is not available within a period of twenty-four hours after the person has been arrested by or delivered to the peace officer, the person shall be taken before a justice as soon as possible,
    unless, at any time before the expiration of the time prescribed in paragraph (a) or (b) for taking the person before a justice,
(c) the peace officer or officer in charge releases the person under any other provision of this Part, or

(d) the peace officer or officer in charge is satisfied that the person should be released from custody, whether unconditionally under subsection (4) or otherwise conditionally or unconditionally, and so releases him.

(2) Conditional release—If a peace officer or an officer in charge is satisfied that a person described in subsection (1) should be released from custody conditionally, the officer may, unless the person is detained in custody for an offence mentioned in section 522, release that person on the person's giving a promise to appear or entering into a recognizance in accordance with paragraphs 498(1)(b) to (d) and subsection 2(1).

(2.1) Undertaking—In addition to the conditions referred to in subsection (2), the peace officer or officer in charge may, in order to release the person, require the person to enter into an undertaking in Form 11.1 in which the person undertakes to do one or more of the following things:

(a) to remain within the territorial jurisdiction specified in the undertaking;
(b) to notify a peace officer or another person mentioned in the undertaking of any change in his or her address, employment or occupation;
(c) to abstain from communicating directly or indirectly, with any victim, witness or other person identified in the undertaking, or from going to a place specified in the undertaking, except in accordance with the conditions specified in the undertaking;
(d) to deposit the person's passport with the peace officer or other person mentioned in the undertaking;
(e) to abstain from possessing a firearm and to surrender any firearm in the possession of the person and any authorization, licence or registration certificate or other document enabling that person to acquire or possess a firearm;
(f) to report at the times specified in the undertaking to a peace officer or other person designated in the undertaking;
(g) to abstain from
   (i)   the consumption of alcohol or other intoxicating substances,
   (ii) the consumption of drugs except in accordance with a medical prescription; and
(h) to comply with any other conditions specified in the undertaking that the officer in charge considers necessary to ensure the safety and security of any victim of or witness to the offence.

(3.1) Interim release—Notwithstanding paragraph (3)(b), a justice may, with the consent of the prosecutor, order that a person referred to in subsection (3), pending the execution of a warrant for the arrest of that person, be released

(a) unconditionally, or

(b) on any of the following terms to which the prosecutor consents, namely,

   (i) giving an undertaking, including an undertaking to appear at a specific time before the court that has jurisdiction with respect to the indictable offence that the person is alleged to have committed, or

   (ii) entering into a recognizance described in any of the paragraphs 515(2)(a) to (e)

with such conditions described in subsection 515(4) as the justice considers desirable and to which the prosecutor consents.

JUDICIAL INTERIM RELEASE

515. (1) Order of release—Subject to this section, where an accused who is charged with an offence other than an offence listed in section 469 is taken before a justice the justice shall, unless a plea of guilty by the accused is accepted, order, in respect of that offence, that the accused be released on his giving an undertaking without conditions, unless the prosecutor, having been given a reasonable opportunity to do so, shows cause, in respect of that offence, why the detention of the accused in custody is justified or why an order under any other provision of this section should be made and where the justice makes an order under any other provision of this section, the order shall refer only to the particular offence for which the accused was taken before the justice.

(2) Release on undertaking with conditions, etc.—Where the justice does not make an order under subsection (1), he shall, unless the prosecutor shows cause why the detention of the accused is justified, order that the accused be released

(a) on his giving an undertaking with such conditions as the justice directs;

(b) on his entering into a recognizance before the justice, without sureties, in such amount and with such conditions, if any, as the justice directs but without deposit of money or other valuable security;

(c) on his entering into a recognizance before the justice with sureties in such amount and with such conditions, if any, as the justice directs but without deposit of money or other valuable security;

(d) with the consent of the prosecutor, on his entering into a recognizance before the justice, without sureties, in such amount and with such conditions, if any, as the justice directs and on his depositing with the justice such sum of money or other valuable security as the justice directs, or

(e) if the accused is not ordinarily resident in the province in which the accused is in custody or does not ordinarily reside within two hundred kilometres of the place in which he is in custody, on his entering into a recognizance before the justice with or without sureties in such amount and with such conditions, if any, as the justice directs, and on his depositing with the justice such sum of money or other valuable security as the justice directs.

(2.1) Power of justice to name sureties in order—Where, pursuant to subsection (2) or any other provision of this Act, a justice, judge or court orders that an accused be released on his entering into a recognizance with sureties, the justice, judge or court may, in the order, name particular persons as sureties.

(2.2) Alternative to physical presence—Where, by this Act, the appearance of an accused is required for the purposes of judicial interim release, the appearance shall be by actual physical attendance of the accused but the justice may, subject to subsection (2.3), allow the accused to appear by means of any suitable telecommunication device, including telephone, that is satisfactory to the justice.

(2.3) Where consent required—The consent of the prosecutor and the accused is required for the purposes of an appearance if the evidence of a witness is to be taken at the appearance and the accused cannot appear by closed-circuit television or any other means that allow the court and the accused to engage in simultaneous visual and oral communication.

(3) Release on undertaking with conditions, etc.—The justice shall not make an order under any of paragraphs (2)(b) to (e) unless the prosecution shows cause why an order under the immediately preceding paragraph should not be made.

(4) Conditions authorized—The justice may direct as conditions under subsection (2) that the accused shall do any one or more of the following things as specified in the order:

(a)   report at times to be stated in the order to a peace officer or other person designated in the order;

(b)   remain within a territorial jurisdiction specified in the order;

(c)   notify the peace officer or other person designated under paragraph (a) of any change in his address or his employment or occupation;

(d)   abstain from communicating, directly or indirectly, with any victim, witness or other person identified in the order, or refrain from going to any place specified in the order, except in accordance with the conditions specified in the order that the justice considers necessary;

(e)   where an accused is the holder of a passport, deposit his passport as specified in the order;

(e.1) comply with any other condition specified in the order that the justice considers necessary to ensure the safety and security of any victim of or witness to the offence; and

(f) comply with such other reasonable conditions specified in the order as the justice considers desirable.

CONCLUSION

This chapter has examined the origins of the law as it relates to interim release in Canada. We have looked briefly at the history of bail and the introduction of the *Bail Reform Act*, which have provided the framework for the current provisions dealing with judicial interim release.

The importance of custody immediately following arrest has been reviewed from the perspective of appropriate police investigative procedures and techniques. Also, this chapter has placed the law of interim release into the context of the *Canadian Charter of Rights and Freedoms*. Background on the impact of pre-trial detention has been considered briefly and the principle of using detention as a "last resort" is acknowledged.

Several cases dealing with specific elements of this subject area have been cited and the most relevant sections of the *Criminal Code* dealing with interim release as it pertains to peace officers, officers in charge, and justices have been reproduced for the benefit of the reader.

QUESTIONS FOR CONSIDERATION AND DISCUSSION

1. In what ways did the *Bail Reform Act* change the practice of pre-trial release of accused persons in Canada?
2. How is the pre-trial, or interim, release of accused persons impacted by the *Canadian Charter of Rights and Freedoms*?
3. How is the concept of "public interest" related to the notion of the "protection or safety of the public"?
4. How does the Canadian criminal law procedure with respect to pre-trial release differ from that in other countries (e.g., the United States, Britain, or Australia)?
5. What circumstances might permit the pre-trial release of someone accused of a serious offence, for example, first degree murder?

CASES CITED

Bietel v. Ouseley (1921), 35 C.C.C. 386 (Sask. C.A.)

R. v. Andrews (1975), 34 C.R.N.S. 344 (Nfld. S.C.)

R. v. Demyen (1975), 26 C.C.C. (2d) 324 (Sask. C.A.)

R. v. Gottfriedson (1906), 10 C.C.C. 239 (B.C. Co. Ct.)

R. v. Kingswatsiak (1976), 31 C.C.C. (2d) 213 (N.W.T.C.A.)

R. v. Lamothe (1990), 58 C.C.C. (3d) 530 (Que. C.A.)

R. v. Phillips (1947), 32 Cr. App. Rep. 47 (C.C.A.)

R. v. Sandhu (1984), 38 C.R. (3d) 56 (Que. S.C.)

Re Powers and the Queen (1972), 9 C.C.C. (2d) 533 (Ont. H.C.)

REFERENCES

Armour, E. (1927). "Annotation: bail in criminal cases." 47 C.C.C. 1

Brown, D.H. (1989). *The genesis of the Canadian Criminal Code of 1892*. Toronto: The Osgoode Society.

Canadian Committee on Corrections. (1969). *Toward unity: criminal justice and corrections: report of the Canadian Committee on Corrections*. Ottawa: Queen's Printer.

Cohen, S. and P. Healy (1982). "A technical note on subsection 454 (1.1) of the Criminal Code and the release powers of peace officers." 24 *Criminal Law Quarterly* 489.

Delisle, Ronald Joseph and Don Stuart (1996). *Learning Canadian criminal procedure*. 4th ed. Toronto: Thomson Canada.

Egan, J.B. (1959). "Bail in criminal cases." *Criminal Law Review*, p. 705.

Friedland, M.L. (1965). *Detention before trial: a study of cases tried in the Toronto Magistrates' Courts*. Toronto: University of Toronto Press.

Friedland, M.L. (1984). *A century of criminal justice: perspectives on the development of Canadian law*. Toronto: Carswell.

Koza, P. and A.N. Doob (1977). "Police attitudes towards the Bail Reform Act." *Criminal Law Quarterly*, Vol. 19, p. 405.

Law Reform Commission of Canada. (1988). *Compelling appearance, interim release and pre-trial detention*. Ottawa: The Commission.

Phelps, M. (1971). "The legal basis of the right to bail." *Manitoba Law Journal*, Vol. 4, p. 143.

McWilliams, P.K. (1967). "The law of bail." *Criminal Law Quarterly*, Vol. 9, p. 21.

Mewett, A.W. (1967). "The criminal law, 1867-1967." *Canadian Bar Review*, Vol. 45, p. 726.

Parker, Graham (1981). "The origins of the Canadian Criminal Code." In Flaherty, D.H. (ed.) *Essays in the history of Canadian law*. Toronto: The Osgoode Society.

Scollin, John A. (1977). *Pre-trial release: being a second edition of the Bail Reform Act: an analysis of provisions of the Criminal Code related to bail and arrest*. Toronto: Carswell.

Trotter, Gary T. (1999). *The law of bail in Canada*. 2nd ed. Scarborough, Ont.: Carswell.

RELATED ACTIVITIES

- Review the fact elements of the cases cited in this chapter and discuss their relevance to the topics of bail in Canada.
- Examine recent issues of *Canadian Criminal Cases*, *Criminal Reports*, or other similar reporting services, for current cases dealing with bail and pre-trial, or interim, release.

- Communicate with local police departments in order to receive background on their policies and procedures dealing with the application of Part XVI of the *Criminal Code*.
- Compare Canada's approach to pre-trial release with similar criminal procedures in other countries (e.g., the United States, Britain, or Australia).

WEBLINKS

 http://www.csc-scc.gc.ca/text/forum/bprisons/english/enge.html A summary of the modern British experience of trying to divert alleged offenders from prison in the pre-adjudication stage.

 http://www.crime-prevention.org/english/publications/youth/mobilize/rrro_e.html This paper describes an innovative program that deals with the large number of young people who cannot raise bail and therefore are remanded in custody until their court date.

 http://www.attorneygeneral.jus.gov.on.ca/html/CRIMJR/execsum-meng.htm#sbail This document examines the operation of the criminal courts in Ontario and, in Chapter 2, makes practical recommendations on pre-trial hearings to increase efficiency.

 http://www.canada.justice.gc.ca/en/dept/pub/hpcp/part2.html#release Police guidelines for criminal harassment cases with sections on "Release from Custody" (Part 2.11) and "Pre-trial Release" (Part 4.4).

CHAPTER FIVE

POLICE DISCRETION IN CANADA

LEARNING OBJECTIVES

1. Describe the concept of discretion as it applies to police officers in Canada.
2. Apply your understanding of police discretion to the executive level of police organizations.
3. Provide reasons to support the need for discretion in the application of police powers in Canada.
4. Distinguish between legitimate and illegitimate applications of police discretion.
5. Describe the characteristics of a training and development approach that will promote effective police discretion.
6. Explain the reasonable limits to police discretion in Canadian law enforcement.
7. Identify the ethical dimension of police discretion.

INTRODUCTION

The topic of police discretion is extremely pertinent to learning about the application of police powers in Canada. Police discretion relates to the dimension beyond where the laws have absolute sway over the specific behaviour of individual officers. Discretion relates to the grey areas that inevitably exist in the broadest range of human actions and reveals itself in the split-second decision-making that is frequently required of police officers as they provide their services to the public. "Police discretion" is the term that represents the critical faculty that individual officers must possess that will allow them to differentiate and discriminate between those circumstances that require absolute adherence to the letter of the law and those occasions when a degree of latitude is justified, based on the officer's knowledge, experience, or instinct.

This chapter will look at the literature that has developed around the issue of police discretion. We will see that discretion is an essential ingredient in modern policing. It will be seen that discretion tends to place a significant reliance on the critical judgment of front-line officers as they function in an environment that is exceedingly self-directed. The criticism of police discretion will be considered and assessed in the context of Canadian law enforcement. Also, the importance of understanding discretion from an ethical perspective will be discussed, particularly as it pertains to the training and development of police recruits.

WHAT IS DISCRETION?

It is useful to begin with a basic definition of what is meant by discretion. Here it is helpful to refer to Ericson (1982):

> Discretion is the power to decide which rules apply to a given situation and whether or not to apply them. Legal scholars traditionally view discretion in terms of what *official* rules can be held to govern the actions of policemen. These rules include laws and administrative instructions. (pp. 11–12) [emphasis in original]

It is important to emphasize the "power" relationship that exists between a police officer and any member of the public. The police retain, in all of their interactions, the capacity to direct encounters according to their own wishes, even against the wishes of the citizens they are dealing with (Ericson, 1982, p. 12).

Delattre (1989) describes discretion in the following manner:

> Discretion is the authority to make decisions of policy and practice. In policing, discretion often includes command or patrol authority to decide which laws shall be enforced, and when, where, and how. It also includes authority to decide which means of helping the helpless, maintaining order, and keeping the peace are best suited to particular circumstances. Discretion is a special kind of liberty—the freedom to make decisions that affect the lives of others, which other citizens are not empowered to make. Special liberties entail special duties. (p. 45)

Delattre goes on to point out that police officers must be permitted a high degree of discretion because it is essentially impossible to establish a regime of laws and regulations that could adequately encompass appropriate behaviour or action in every conceivable circumstance.

POLICE DISCRETION IN THE UNITED STATES AND CANADA

In the United States, the topic of police discretion was addressed with some degree of seriousness by the 1967 President's Commission on Law Enforcement and the Administration of Justice. This was a comprehensive review of policing in the U.S. that began a period of debate on the appropriate role of discretion. It has been noted

(Guyot, 1991) that by admitting to the existence of wide police officer discretion several uncomfortable questions arise. For example:

- How can the overall fairness of law enforcement be preserved if police services are not centrally controlled?
- What skills, aptitudes, abilities, and attitudes do police officers require in order to apply discretion wisely?

A number of authors have highlighted the willingness of police administrators to sustain the fiction that discretion does not in fact guide a substantial amount of police activity (Grosman, 1975; Goldstein, 1985; Sheehan & Cordner, 1989; and Guyot, 1991). It is valuable, from the point of view of public confidence, to ensure that the police are seen as an impartial agency of the state, and one that dispenses their portion of the criminal justice system in an impartial and balanced manner. The public generally assumes that police officers are engaged in what may be referred to as full enforcement. That is to say, all laws and regulations are fully and equally applied by the police, regardless of circumstances. It is typically believed that the courts make the fine distinctions on specific points of law, including mitigating factors and special considerations raised by either Crown counsel or defence lawyers. There is a common view that police officers follow substantial rules of conduct, apply massive codes of criminal law, and adhere to staggering internal policies and procedures. To some extent, this view is correct; however, it remains the case that the individual officer, functioning as a front-line service provider, has incredible latitude in the application of those rules, codes, policies, and procedures.

In Canada, police discretion has followed much of the pattern established in the United States. It may be put forward as a general principle that it is virtually impossible to provide rules and regulations for police officer activity in the field that would eliminate the need for officers to use their discretion. Instead, officers must be given a significant margin of flexibility to make independent assessments of situations and to choose whether or not to invoke their law enforcement or order-maintenance powers.

DISCRETION AND CONSIDERATIONS FOR POLICE TRAINING AND LEARNING

Indeed, recruit training in many police departments places a heavy emphasis on the need for officers to adhere to the rules and regulations of the organization. They are instructed on the law and the application of the statutes dealing with law enforcement, order maintenance, and emergency response. A great deal of attention is paid to the whole structure of controls that channel police powers and the front-line officer's responsibility to her or his supervisor (e.g., shift sergeant). However, front-line officers, by the very nature of police work, are frequently alone in the performance of their sworn duties. This makes it impossible for the supervisor to be available for every decision that will be required of the officers as they go about their patrols. While the existence of many rules and regulations may serve to guide front-line officers in the performance of their duties, there will always be considerable scope for the officer's individual discretion based on the multitude of circumstances that she or he will confront on a routine basis. The contrast between the rather academic business of learning about the rules and regulations that provide the framework for police powers in Canada and the actual street reality of policing is considerable. The responsibility of knowing the law and the limits of police authority is one thing, the capacity to take that knowledge and apply it to the real world of day-to-day law enforcement is quite another. This is where discretion enters the equation. It rests on an ability to make reasonable decisions under frequently difficult circumstances. As Wilson (1978) notes from his in-depth study of police behaviour:

> The patrolman's decision whether and how to intervene in a situation depends on his evaluation of the costs and benefits of various kinds of action. Though the substantive criminal law seems to imply a mandate, based on duty or morality, that the law be applied wherever and whenever its injunctions have been violated, in fact for most officers there are considerations of utility that equal or exceed in importance those of duty or morality, especially for the more common and less serious laws. (pp. 83–84)

Guyot (1991, pp. 41–50) includes the following attributes for police officers who will be exercising discretion:

- *Curiosity*—officers need this quality in order to pursue important questions that will allow them to determine if police intervention is necessary;
- *Judgment of danger*—this is a critical capacity that allows officers to make skilful assessments about the dangerousness of a situation in which they have decided to intervene;
- *Tragic perspective*—Muir (1977) coined this term that relates to an understanding of people and human nature. It speaks to the role of compassion in policing and draws extensively upon the officer's life experiences;
- *Decisiveness*—this relates to an ability to apply necessary coercion in a crisis without undue delay, and to know when such direct and immediate action is necessary; and
- *Self-control*—this essential personal quality is needed in dealing effectively with others who may be in the grip of high emotion. It relates to composure under extreme pressure.

In many instances, policing training programs should concentrate their attention on what police officers should *not* do in the execution of their duties. It is often valuable to proscribe certain kinds of action or behaviour when detailing particularly sensitive areas of police activity.

Wilson (1978) provides insight into the continuing need for police officers to be trained in the appropriate techniques of intervention that will allow them to function competently once their discretionary capabilities have been exercised. Police officers should be highly trained in various modes for approaching a suspect or violator, and in what approaches to take when dealing with people with mental disabilities, or others who may require special handling. Specifically he notes:

> With respect to his preventive patrol function, the patrolman can be given a clear statement about how to intervene even if not about whether to intervene. He can be taught how to approach a suspect, what to say, what kind of identification to ask for, what other questions to put, and how to check the name by radio to see if the person is wanted. (p. 65)

Wilson (1978) believes that it is important that officers be given as much assistance as possible when it comes to the application of discretion:

> Some thoughtful observers of police practices have suggested that the strain can be reduced if the patrolmen are given clearer substantive guides to the use of their discretion. To the extent this is possible it is of course desirable. At the very least, certain obvious steps can be taken once the fiction that the police have no discretion is dropped. (pp. 293–294)

Goldstein (1985) offers some forceful observations on the nature of police recruit training in the U.S. context. He asserts that:

> Recruit training in police agencies is frequently inadequate because the instruction bears little relationship to what is expected of the officer when working in the field. In the absence of guidelines that relate to an analysis of police experience, the instructor is usually left with only the formal definition of police authority to communicate to the trainee, and this is often translated to the student merely by the reading of statutory definitions. Students are taught that all laws are to be fully enforced. The exercise of police authority is similarly taught in a doctrinaire fashion. With this kind of formal training, the new officer finds, upon assignment to the field, the necessity of acquiring from the more experienced officers a knowledge of all the patterns of accommodations and modifications. As an awareness emerges of the impracticality and lack of realism of much of what was learned as a student, the recruit unfortunately begins to question the validity of all aspects of formal training. (p. 53)

It is presumed that most police recruits will arrive for initial training without a great deal of background in problem-solving or decision-making (Wilson, 1978, p. 88). However, in the context of police learning in Ontario this situation is being addressed in many innovative ways. For example, the Ontario Strategic Planning Committee on Police Training and Education undertook in-depth research to discover that these abilities will be critical to police officers (at all rank levels) in the future

(Ontario. Ministry of the Solicitor General. Strategic Planning Committee on Police Training and Education, 1992).

It is imperative that those in charge of police academies recognize the enormous responsibility that falls upon police officers who function in the front lines of law enforcement. Accordingly, it is their task to design training and learning programs that appropriately empower officers to apply decision-making and discretionary skills. As Wilson (1978) indicates:

> The lack of any guidelines for police discretionary action, while its exercise is widely utilized, places the decision-making burden squarely upon the shoulders of individual policemen. Some acknowledgement and control is required. The open recognition within the police academy that policemen do not just follow rules should eventually result in a consideration of how to exercise that discretion without violating the rule of law. (p. 93)

Training and development programs must guard against the danger of allowing officers to fall into the trap of stereotyping when they are dealing with the public. If there are no hard-and-fast rules for specific law enforcement or order-maintenance activities, officers may be tempted to rely on convenient, though mistaken, stereotypes when dealing with certain categories of people.

THE PARADOX OF POLICE DISCRETION

There is, clearly, a paradox that exists within modern policing, as it relates to police discretion, that creates significant difficulties for police officers and senior administrators alike. How can police organizations address the absolute necessity for substantial discretion on the part of their officers, particularly those in front-line service delivery, while continuing to maintain public trust, confidence, and accountability? If an individual officer is accorded a wide margin of discretion with respect to the application of the law, how is the public going to be assured that this discretion will not be abused or misapplied? If the individual officer has such a high degree of discretion in the day-to-day application of the law, what measure of discretion is being applied by the senior executives of the police organization whose functions are generally hidden from public scrutiny? These, and other related questions, must be addressed in the context of policing in a democratic society. The impartial application of the law is one of the hallmarks of democratic rule and any departure from such an application is seen as being arbitrary and inconsistent with our principles of fundamental justice. Grosman (1975) makes the following observation about the impact of discretion on policing:

> If individual police officers exercise substantial power by the decisions they make which determine the future of individuals, what about the policy decisions taken by their commanding officers? If the criteria used by individual police officers in their exercise of discretion are for

the most part unknown, how much more mysterious and covert are the criteria upon which internal policy decisions and commands of the Chief of Police depend? (p. 3)

The challenge of police discretion is one that has been difficult to address with complete candour. Police executives want to assure the public that they are in charge of fully accountable, responsible, and regulated organizations. However, they know from experience that their officers must have considerable latitude in whether or not they intervene in specific instances. Sheehan and Cordner (1989) speak to the ambiguity that this dilemma fosters:

Police executives typically deny the existence of discretion, partly because they find it difficult to justify and partly because admitting its existence might bring demands for accountability in its control. By establishing and enforcing great numbers of rules and regulations, police executives create the impression that police officers work under severely controlled conditions. On close examination, however, these rules can be seen to affect peripheral matters, such as when the uniform hat must be worn or whose approval of overtime is required, rather than the actual performance of police duties. (pp. 78–78)

Guyot (1991) indicates the following:

However much discretion police officers actually exercise, the police field has ignored, denied, and discouraged discretion. Instead, it has made control the paramount issue throughout the history of American policing. Certainly departmental rules and regulations are essential for such fundamental matters as restricting the use of deadly force and forbidding the acceptance of gratuities. They establish procedures for routine actions, set the limits on discretion, and lend formal recognition to priorities, but they serve poorly as the major means for controlling most actions. (p. 59)

Sheehan and Cordner (1989) offer some guidance on overcoming this paradox of police discretion for individual officers:

The myth of full enforcement and the reality of discretion create ethical dilemmas for the police. Most police officers promise to uphold the law when they are sworn to duty, and the laws of many states mandate that officers shall arrest whenever they have sufficient evidence. Officers must recognize the legislative intent behind such language, in order to avoid feeling that they are shirking their duties when they exercise their discretion not to enforce the law. To assist officers in resolving this problem, police codes of ethics should emphasize service to the public and commitment to the primary goals of protection of life and property and maintenance of order, rather than emphasizing strict enforcement of the law. (pp. 53–54)

Further compounding the original paradox of the necessity of police discretion in the face of organizational denials of its existence is the additional irony that those at the bottom rank level of the police hierarchy are the ones most heavily endowed with discretion. As Grosman (1975) observes:

We must now explore the discretionary input of the individual police officer: for unlike other bureaucratic organizations the police force gives to its lowest-ranking members the power to make critical decisions in making or not making arrests. A great deal of the discretion that he exercises would be considered by many as illegal or, at best, of questionable legality. (p. 81)

It is obvious from an observation of front-line police officers in action that discretion is part of their working life. They are constantly sizing up situations and making decisions about pursuing or not pursuing the letter of the law. The myth of full enforcement has been described by Pratt (1985):

> The simple fact is, police officers must make and do make literally millions of discretionary decisions every day, all across this land. We could not operate otherwise. A policy of full enforcement is both undesirable and unachievable. (p. 65)

A Canadian perspective on the myth of full enforcement is offered by Grosman (1975) in his study of police executives:

> The police themselves have, until recently, subscribed publicly to the myth of full enforcement. It is only lately that senior police administrators have acknowledged the need for the continual exercise of their discretion. (p. 84)

The case of *R. v. Commissioner of Police of the Metropolis, ex parte Blackburn and Another (No. 3)* (1972), 2 W.L.R. 43 provides an instance where a senior police executive openly spoke to the issue of full enforcement and the fact that it did not occur.

Grosman (1975) compares the role of the police officer acting in his or her discretionary capacity with the trial judge:

> The decision whether to invoke the awesome machinery of the criminal justice system is often made by the policeman on the street. He is the one who decides whether or not to arrest an individual. This single crucial decision seriously damages the future of an individual, and may determine whether or not he is ever convicted of any criminal charge. If the police officer decides not to arrest, he may permit an offender the opportunity to commit further crimes. In this way the officer may ultimately exercise greater discretion over the individual and his future than does the judge himself. (p. 1)

HOW DOES POLICE DISCRETION MANIFEST ITSELF IN PRACTICE?

We have looked at an appropriate definition of police discretion and have examined the necessity of this capacity within the police function. But how does this concept translate into practice? Are there specific steps that an officer will take in order to apply his or her discretion, or is it something that defies description? Discretion will reveal itself across the entire spectrum of police activity as Guyot (1991) has noted:

> To discharge varied and demanding responsibilities requires an officer first to decide whether to look into the situation, and then to decide what to do. The term "exercising discretion" has

long been understood in policing to apply to situations where an officer has legal authority to make an arrest or issue a ticket but decides not to do so. This definition is too narrow, because independent choice and action permeate every aspect of patrol work. Recently, the term "discretion" has acquired the broader meaning that includes all independent decisions made by line officers. (p. 41)

In a very useful summary, Sheehan and Cordner (1989) have enumerated the following reasons for police officers exercising their discretion in patrol situations:

1. a police officer who attempted to enforce all the laws all the time would be in the station house and in court all the time and, thus, of little use when problems arose in the community;
2. legislatures pass some laws that they clearly do not intend to have strictly enforced all the time;
3. legislatures pass some laws that are vague, making it necessary for the police to interpret them and decide when to apply them;
4. most law violations are minor in nature (speeding one mile per hour over the limit, parking 13 inches from the curb) and do not require full enforcement;
5. full enforcement of all the laws all the time would alienate the public and undermine support for the police and the legal system;
6. full enforcement of all laws all the time would overwhelm the courts and the correctional system; and
7. the police have many duties to perform with limited resources; good judgment must, therefore, be used in the establishment of enforcement priorities. (pp. 52–53)

As part of his major study of police behaviour, Wilson (1978, p. 84) developed a chart that can assist us in understanding the basis for applying discretion. He makes an important distinction between the terms "law enforcement" and "order maintenance." Also, he differentiates actions that are initiated by the police officer from those that are initiated by a citizen. Within the context of these four categories, discretion will arise in distinctive ways.

Wilson (1978, p. 85) describes four sorts of discretionary situations, as outlined in Figure 5.1, below.

In each of these categories, a different degree of discretion will be available for the police officer, the police service, or both.

In Case I, the police commence some form of action with respect to a law enforcement issue, i.e., a matter that involves a violation of the law, or a legal infraction. Police administrators may have some influence on these activities by establishing service-wide policies with respect to traffic offences, gambling, or some other regulated activities.

Case II involves law enforcement activity that is initiated by a citizen, or citizens, who complain to the police officer. In these instances the officer functions in order to take information and gather evidence that may lead to arrest. For example, this category could

FIGURE 5.1

Four Kinds of Discretionary Situations

BASIS OF POLICE RESPONSE

		Police-Invoked Action	Citizen-Invoked Action
NATURE OF SITUATION	Law Enforcement	I	II
	Order Maintenance	III	IV

Source: Wilson, James Q. *Varieties of police behavior: the management of law and order in eight communities*. Cambridge, Mass.: Harvard University Press, p. 85. Reprinted by permission. Copyright © 1968, 1978 by the President and Fellows of Harvard College.

involve instances of shoplifting, auto theft, burglary, or fraud. Police administrators have a degree of control over these cases, particularly if they involve young offenders. It has been accepted as a convention of modern policing that officers are allowed a wide margin of discretion in their dealings with young persons.

In Case III, a police officer will apply discretion to decide on whether or not to intervene in a matter that involves real, or perceived, disorder. Examples of order maintenance include such things as breaches of the peace, public drunkenness, and disorderly conduct. A police administrator may establish some guidelines for front-line officer discretion; however, it would be virtually impossible to monitor absolute compliance with such guidelines in these areas. The patrol officer is in the best position to control discretion in the order-maintenance category.

Finally, Case IV deals with order maintenance concerns that are commenced by a citizen, or citizens. Here someone will call for the police to assist in some private or public instance of disorder. Typically, there *must* be some form of police response to these calls. The police department retains some latitude in determining the nature of the specific response, and it often falls to the individual front-line officer to assess the situation in order to judge which response is appropriate. The nature of the response will hinge on the personal style of the police officer and the characteristics of the citizens involved, rather than on detailed departmental policy.

By way of summary, Wilson (1978) explains that:

> . . . in Cases I and IV the patrolman has great discretion, but in the former instance it can be brought under departmental control and in the latter it cannot. In Case II the patrolman has the least discretion except when the suspects are juveniles and then the discretion is substantial and can be affected by general departmental policies and organization. Case III is intermediate in both the degree of discretion and the possibility of departmental control. (p. 89)

POLICE DISCRETION AS AN EARMARK OF PROFESSIONALISM

In the context of any profession, there is an understanding that its practitioners will be in possession of some authoritative knowledge and skills that allow them to function at a high level of efficiency and effectiveness. Often that knowledge and those skills are acquired through a rigorous course of study at an educational faculty affiliated with the profession. Generally, it is known that most police services in Canada require their recruits to undergo lengthy academic and practical training and education prior to being sworn as peace officers. Furthermore, there are intensive programs of continuous learning that police officers are required to successfully complete if they are to proceed through the ranks, or, if they wish to assume additional responsibilities. In the spirit of professionalism, many departments have attempted to summarize their understanding of police discretion. Sheehan and Cordner (1989), for example, make reference to the standard on police discretion that was advanced under the auspices of the U.S. National Advisory Commission on Criminal Justice Standards and Goals:

STANDARD ON POLICE DISCRETION

Standard 1.3
Police discretion

1. *Every police agency should acknowledge the existence of the broad range of administrative and operational discretion that is exercised by all police agencies and individual officers. That acknowledgment should take the form of comprehensive policy statements that publicly establish the limits of discretion, that provide guidelines for its exercise within those limits, and that eliminate discriminatory enforcement of the law.*

2. *Every police chief executive should establish policy that guides the exercise of discretion by police personnel in using arrest alternatives.*

3. *Every police chief executive should establish policy that limits the exercise of discretion by police personnel in conducting investigations, and that provides guidelines for the exercise of discretion within those limits.*
4. *Every police chief executive should establish policy that governs the exercise of discretion by police personnel in providing routine peacekeeping and other police services that, because of their frequent recurrence, lend themselves to the development of a uniform agency response.*

Source: National Advisory Commission on Criminal Justice Standards and Goals, *Police* (Washington, D.C.: U.S. Government Printing Office, 1973), pp. 21–22.

From a Canadian perspective with regard to the broad framework of police discretion the following statement of principles provides a preamble to the *Police Services Act* in Ontario:

1. The need to ensure the safety and security of all persons and property in Ontario;
2. The importance of safeguarding the fundamental rights guaranteed by the Canadian Charter of Rights and Freedoms and the Human Rights Code;
3. The need for co-operation between the providers of police services and the communities they serve;
4. The importance of respect for victims of crime and understanding of their needs;
5. The need for sensitivity to the pluralistic, multiracial and multicultural character of Ontario society; and
6. The need to ensure that police forces are representative of the communities they serve.

While these principles could never be seen as a complete guide to specific officer behaviour, they do provide a valuable framework within which an officer can function. They provide a sense of the corporate culture of policing in Ontario as expressed through these statements. Indeed, they go some distance toward fostering a sense of police officer discretion as they articulate operating principles that can inform action in a variety of instances.

Wilson (1978) makes an observation that places the front-line police officer in a unique position as a "professional." Because the individual officer (or "sub-professional") works alone, there is a high level of trust placed on that person's shoulders that they will acquit themselves appropriately, even in the most challenging situations:

. . . the police administrator is in the same position as the school superintendent, the hospital director, or the secretary of state. But unlike these others, he must deal with the problem of discretion being exercised by sub-professionals who work alone (and thus cannot be constrained by professional norms) and in situations where, both because the stakes are high and the environment apprehensive or hostile, the potential for conflict and violence is great. (p. 67)

POLICE DISCRETION AND COMMUNITY POLICING

The existence of police discretion should also be understood within the context of community policing. McKenna (2000) has provided detailed background on the theory and practice of community policing in Canada; and the manner in which police officers apply their considerable discretion will have an inevitable bearing on the nature and quality of policing delivered in the community context.

As we have seen earlier in this chapter, once the reality of police discretion has been made known to the public, the inevitable questions arise with respect to the guidelines provided to officers for the exercise of discretion. Also, important considerations relevant to police accountability must be addressed and resolved. For example, if the individual police officer has discretion in the application of law enforcement and order-maintenance rules, how can one segment of the community be assured that they are not being unfairly or inappropriately targeted by the police? Will officers be harsher when patrolling poorer communities? Will officers apply their discretion to turn a "blind eye" in communities that include influential citizens? These are serious questions that speak to the professional and ethical concerns relevant to police discretion.

Credit: Mike Weaver, Media Relations, Kingston Police

Police discretion takes many forms in Canada.

The reality that police discretion creates conditions whereby different approaches to patrol can occur is touched on by Wilson (1978):

> Though the legal and organizational constraints under which the police work are everywhere the same or nearly so, police behavior differs from community to community. (p. 83)

Wilson (1978) also notes that communities do not fully comprehend the extent to which police discretion will override their own community policing choices. In spite of all the rules and regulations in place, individual police behaviour is substantially guided by the officer's own values, attitudes, and professional character. Wilson (1978) asserts that:

> . . . deliberate community choices rarely have no more than a limited effect on police behavior, though they may often have a great effect on police personnel, budgets, pay levels, and organization. How the police, especially the patrolmen, handle the routine situations that bring them most frequently into contact with the public can be determined by explicit political decisions only to the extent that such behavior can be determined by the explicit decisions of the police administrator . . . [T]he administrator's ability to control the discretion of his subordinates is in many cases quite limited by the nature of the situation and the legal constraints that govern police behavior. (p. 227)

Wilson (1978) makes the valuable observation that communities that share certain values will have a much higher level of comfort with police discretion than those where common ground is not shared:

> The problems created by the exercise of necessary discretion are least in communities that have widely shared values as to what constitutes an appropriate level of order and what kind of person or form of behavior is an empirically sound predictor of criminal intentions: the problems are greatest in cities deeply divided along lines of class and race. (pp. 278–279)

However, the fact that there may be high tolerance within any given community for substantial police discretion does not mean that police administrators should abdicate their responsibility to set out benchmarks for police activity within the community. Wilson (1978) offers the following insight:

> To say that discretion inevitably exists and thus cannot be reduced to rule is not to say that the police administrator does not know what he wishes done. Often, perhaps always, he will believe that there is a right and a wrong way to handle the situation and that if he were on the spot he would know what to do . . . But he cannot in advance predict what the circumstances are likely to be or what courses of action are most appropriate—because, in short, he cannot be there himself—he cannot in advance formulate a policy that will guide the patrolman's discretion by, in effect, eliminating it. (p. 66)

In many instances, the police organization is compelled to make certain discretionary decisions that move them away from the mythic ideal of full enforcement, as well as, from possible public safety and order goals advanced within the community. Wilson (1978) observed in his study of eight U.S. communities that municipal au-

A police officer gets to know some children in the comfortable setting of the classroom.

Credit: Mike Weaver, Media Relations, Kingston Police

thorities make financial decisions that create circumstances whereby the police must prioritize their operational activities:

> Limited municipal resources for political purposes inevitably result in the establishment of priorities for law enforcement. The discretion is left with the police to use their resources on the basis of a set of priorities which are at best ill defined. There are no guidelines or principles for the exercise of such discretion in the enforcement of some laws and the non-enforcement of others. (p. 85)

It is vitally important that police organizations comprehend the impact of their use of discretion as it relates to the goals of community policing. Inevitably, comparisons will be made across different communities within the police service's jurisdiction and the police executives must be prepared to justify any variations in their application of such discretion. As Wilson (1978) notes:

> This distinct exercise of discretion by police officers determines the kind of persons who enter our criminal justice system. This grey area of decision-making is unknown to the public who, for the most part, continue to believe that the law is enforced as it is written—that any deviation by the police from full enforcement is unlawful. (p. 81)

Wilson further notes:

> Since there are no established criteria for uniform action in the exercise of discretion, an acknowledgement of discretionary decisions opens police administration to the criticism that they do not adhere to uniform procedures and that they deviate from legislative instructions. (p. 84)

Goldstein (1985) notes:

> Another popular misconception is that the police are a ministerial agency, having no discretion in the exercise of their authority. While this view is occasionally reinforced by a court decision, there is a growing body of literature that cites the degree to which the police are, in fact, required to exercise discretion—such as in deciding which laws to enforce, in selecting from among available techniques for investigating crime, in deciding when to arrest, and in determining how to process a criminal offender. (p. 45)

From a community policing perspective, it is worthwhile to consider the establishment of criteria that will guide the application of police discretion. Whatever contributes to predictability and consistency in the use of police powers is a benefit. This point is articulated by Goldstein (1985) when he explains that:

> In the long run, the exercise of discretion in accordance with defensible criteria would create greater confidence in the police establishment. More immediately, it would lead to a reduction in the number of arbitrary actions taken by individual officers, thereby substantially reducing the tensions which such actions often create—particularly in areas in which minority groups are affected. (p. 51)

THE ETHICAL DIMENSION OF DISCRETION

While this chapter cannot do justice to the topic of police ethics, it is relevant that there be some linkage between the topic of police discretion and that of ethical police conduct. Recognizing that police officers assume a significant measure of discretion in the exercise of their enormous powers leads to the inevitable speculation as to whether or not these officers have the moral, critical, and professional capacity to acquit themselves of this responsibility. If front-line police officers function to such a high degree on their own, with minimal supervision and little direct oversight, what proof does the average citizen have that there will be fairness, justice, and integrity evident in all interactions with the police? There is a dilemma implicit in the police administrator's public disavowal of police discretion when it comes to public complaints.

Wilson (1978) highlights the concern that may arise with respect to public complaints and the possibility that a resolution may seem to be "political":

> Because he works alone his supervisor can never know exactly what happened and must take either his word or the complainant's. The patrolman necessarily exercises wide discretion, but the police administrator is obliged publicly to deny that there is much discretion in police work and may have to act out that denial by deciding complaints about police behavior as if no discretion existed. (p. 72)

> It is not that the patrolman is frequently punished for misusing his discretionary powers but that such punishment as occurs (being based on ambiguous rules applied after the fact in situations where the citizen's word is taken as against the officer's) is experienced by him as arbitrary or the result of "political influence". (p. 73)

In order to combat the potential for corruption, Delattre (1989) recommends that police departments publish forthright policies with respect to discretion. Police officers that are imbued with high moral and ethical standards, combined with a full array of technical capabilities and capacities, will advance the highest goals of the organization:

> As with enforcement, truthful policies within which officers can responsibly exercise discretion are a basic obligation of command. Fulfillment of this obligation enables openness and obstructs corruption by increasing the possibility of acting on the words the department says it lives by. (p. 50)

CONCLUSION

This chapter has provided the reader with an overview of the subject of police discretion. We have considered the meaning of police discretion and how it has come to be seen as an absolute necessity for modern policing in Canada, and elsewhere. The fact that no police organization could ever hope to provide sufficient policies and procedures to control every conceivable situation is acknowledged.

We have reviewed the essential paradox that makes it very difficult for police executives to "confess" to the degree of discretion allowed to front-line patrol officers, and the fact that full enforcement is, essentially, a myth. However, this healthy confession requires that the training, education, and development of police officers (particularly new recruits) must be of the highest calibre in order to provide them with an appropriate understanding of the principles and qualities that ought to guide their operational behaviour.

This chapter has looked at the relationship between police discretion and community policing. It may be suggested that the higher the degree of officer discretion, the greater the degree of emphasis that must be placed on the accountability mechanisms that will ensure that community wellness is being advanced by the police organization as a whole. We have considered the professional nature of policing and have attempted to understand that police discretion plays a critical role in the ongoing professionalization of policing in Canada.

QUESTIONS FOR CONSIDERATION AND DISCUSSION

1. How can police officers develop curiosity about individuals and patterns of behaviour?
2. How can police officers develop the ability to make correct judgments about danger?
3. How can police officers cultivate a compassionate understanding of the people they serve?
4. How can police officers develop decisiveness and come to terms with the necessity to use coercion?
5. How can police officers learn to exercise significant levels of self-control?

REFERENCES

Bolt, E. et al. (1977). *Police policymaking: the structuring of discretion in the use of criminal investigation procedures: final report, phase I*. Boston: Boston Police Department.

Breitel, Charles D. (1960). "Controls in criminal law enforcement." *University of Chicago Law Review*, Vol. 27 (Spring), p. 427.

Brenner, Robert N. and Marjorie Kravitz (1978). *Police discretion: a selected bibliography*. Washington, D.C.: National Institute of Law Enforcement and Criminal Justice.

Coleman, Stephen F. (1986). *Street cops*. Salem, Wis.: Sheffield Publishing Co.

Davis, Kenneth Culp (1969). *Discretionary justice*. Baton Rouge, La.: Louisiana State University Press.

Davis, Kenneth Culp (1975). *Police discretion*. St. Paul, Minn.: West Publishing.

Delattre, Edwin J. (1989). *Character and cops: ethics in policing*. Washington, D.C.: American Enterprise Institute for Public Policy Research.

Ericson, Richard V. (1982). *Reproducing order: a study of police patrol work*. Toronto: University of Toronto Press in association with the Centre of Criminology.

Finckenhauer, J.O. (1976). "Some factors in police discretion and decision-making." *Journal of Criminal Justice*, Vol. 4, no. 1 (Spring), pp. 29–46.

Goldstein, Herman (1963). "Police discretion: the ideal versus the real." *Public Administration Review*, Vol. 23 (September), pp. 140–148.

Goldstein, Herman (1967). "Administrative problems in controlling the exercise of police authority." *Journal of Criminal Law, Criminology, and Police Science*, Vol. 58, no. 2 (June), pp. 160–172.

Goldstein, Herman (1977). *Policing in a free society*. Cambridge, Mass.: Ballinger Publishing Co.

Goldstein, Herman (1985). "Police policy formulation: a proposal for improving police performance." In More, Harry W. (ed.) *Critical issues in law enforcement*. 4th ed. Cincinnati, Ohio: Anderson Publishing Co.

Goldstein, Joseph (1960). "Police discretion not to invoke the criminal process: low-visibility decisions in the administration of justice." *Yale Law Review*, Vol. 69 (March), pp. 593–594.

Grosman, Brian A. (1975). *Police command: decisions & discretion*. Toronto: Macmillan of Canada.

Guyot, Dorothy (1991). *Policing as though people matter*. Philadelphia: Temple University Press.

McKenna, Paul F. (2000). *Foundations of community policing in Canada*. Scarborough, Ont.: Pearson Education Canada.

Muir, William Ker (1977). *Police: streetcorner politicians*. Chicago: University of Chicago Press.

Ontario. Ministry of the Solicitor General. Strategic Planning Committee on Police Training and Education. (1992). *Report on strategic learning requirements for police personnel*. Toronto: The Ministry.

Petersen, David Muir (1969). The police, discretion and the decision to arrest [Ph.D. dissertation, University of Kentucky]. Ann Arbor, Mich.: University Microfilms.

Pound, R. (1960). "Discretion, dispensation and mitigation: the problem of the individual special case." *New York University Law Review*, Vol. 35, p. 925.

Pratt, C.E. (1985). "Discretion: the essence of professionalism." In More, Harry W. (ed.) *Critical issues in law enforcement*. 4th ed. Cincinnati, Ohio: Anderson Publishing Co.

Reiss, A.J. (1970). "Discretionary decision making and professionalization." In Chapman, Samuel G. (ed.) *Police patrol readings*. 2nd ed. Springfield, Ill.: Charles C. Thomas.

Reiss, A.J. (1974). "Discretionary justice." In Glaser, D. (ed.) *Handbook of criminology*. Chicago, Ill.: Rand McNally.

Remington, Frank J. (1965). "The role of the police in a democratic society." *Journal of Criminal Law, Criminology and Police Science*, Vol. 56, pp. 361–365.

Sheehan, Robert and Gary W. Cordner (1989). *Introduction to police administration*. 2nd ed. Cincinnati, Ohio: Anderson Publishing Co.

Steer, D. (1970). *Police cautions: a study in the exercise of police discretion*. London: Basil Blackwell.

Thibault, Edward A., Lawrence M. Lynch, and R. Bruce McBride (1985). *Proactive police management*. Englewood Cliffs, N.J.: Prentice-Hall.

U.S. President's Commission on Law Enforcement and the Administration of Justice (1967). *The challenge of crime in a free society*. Washington, D.C.: Government Printing Office.

Vorenberg, J. (1976). "Narrowing the discretion of criminal justice officials." *Duke University Law Journal*, (September), no. 4, pp. 651–697.

Wilson, James Q. (1978). *Varieties of police behavior: the management of law and order in eight communities*. Cambridge, Mass.: Harvard University Press.

RELATED ACTIVITIES

- Examine local newspaper articles dealing with the application of police discretion (e.g., the decision of the RCMP to deploy pepper spray against demonstrators at the APEC conference in Vancouver, B.C.).

- Research your local police service. Examine policies that may exist within different police jurisdictions with regard to police discretion. Have any guidelines or directives been established for the exercise of discretion by individual officers?
- Review complaints against individual officers from a given jurisdiction in order to determine if there is a relationship between the application of police discretion and public complaints.

WEBLINKS

 http://reseau.chebucto.ns.ca/Law/LRC/policeV.html At this site, the section titled "Police Protocols" discusses police discretion in the context of spousal abuse.

 http://www.sgc.gc.ca.epub/pol/e199802/e199802.htm This page contains a report to the Solicitor General of Canada detailing influences on police judgment in young offender cases.

 http://www.rcmp-learning.org/docs/ecdd1222.htm In this article, the authors discuss how the transition from "honest cop" to "compromised officer" can occur and how ethics can be taught and maintained in law enforcement agencies.

APPENDIX

CANADA ACT 1982
including the
CONSTITUTION ACT, 1982
1982, c. 11 (U.K.)
[29th March 1982]

[Note: The English version of the Canada Act 1982 is contained in the body of the Act; its French version is found in Schedule A. Schedule B contains the English and French versions of the *Constitution Act, 1982.*]

An Act to give effect to a request by the Senate and House of Commons of Canada.

Whereas Canada has requested and consented to the enactment of an Act of the Parliament of the United Kingdom to give effect to the provisions hereinafter set forth and the Senate and the House of Commons of Canada in Parliament assembled have submitted an address to Her Majesty requesting that Her Majesty may graciously be pleased to cause a Bill to be laid before the Parliament of the United Kingdom for that purpose.

Be it therefore enacted by the Queen's Most Excellent Majesty, by and with the advice and consent of the Lords Spiritual and Temporal, and Commons, in this present Parliament assembled, and by the authority of the same, as follows:

1. *The Constitution Act, 1982* set out in Schedule B to this Act is hereby enacted for and shall have the force of law in Canada and shall come into force as provided in that Act. — Constitution Act, 1982 enacted

2. No Act of the Parliament of the United Kingdom passed after the *Constitution Act, 1982* comes into force shall extend to Canada as part of its law. — Termination of power to legislate for Canada

French version

3. So far as it is not contained in Schedule B, the French version of this Act is set out in Schedule A to this Act and has the same authority in Canada as the English version thereof.

Short title

4. This Act may be cited as the Canada Act, 1982.

SCHEDULE B

CONSTITUTION ACT, 1982

PART I

CANADIAN CHARTER OF RIGHTS AND FREEDOMS

Whereas Canada is founded upon principles that recognize the supremacy of God and the rule of law:

Guarantee of Rights and Freedoms

Rights and freedoms in Canada

1. The Canadian Charter of Rights and Freedoms guarantees the rights and freedoms set out in it subject only to such reasonable limits prescribed by law as can be demonstrably justified in a free and democratic society.

Fundamental Freedoms

Fundamental freedoms

2. Everyone has the following fundamental freedoms:

(*a*) freedom of conscience and religion;
(*b*) freedom of thought, belief, opinion and expression, including freedom of the press and other media of communication;
(*c*) freedom of peaceful assembly; and
(*d*) freedom of association.

Democratic Rights

Democratic rights of citizens

3. Every citizen of Canada has the right to vote in an election of members of the House of Commons or of a legislative assembly and to be qualified for membership therein.

4. (1) No House of Commons and no legislative assembly shall continue for longer than five years from the date fixed for the return of the writs at a general election of its members.

Maximum duration of legislative bodies

(2) In time of real or apprehended war, invasion or insurrection, a House of Commons may be continued by Parliament and a legislative assembly may be continued by the legislature beyond five years if such continuation is not opposed by the votes of more than one-third of the members of the House of Commons or the legislative assembly, as the case may be.

Continuation in special circumstances

5. There shall be a sitting of Parliament and of each legislature at least once every twelve months.

Annual sitting of legislative bodies

Mobility Rights

6. (1) Every citizen of Canada has the right to enter, remain in and leave Canada.

Mobility of citizens

(2) Every citizen of Canada and every person who has the status of a permanent resident of Canada has the right
(a) to move to and take up residence in any province; and
(b) to pursue the gaining of a livelihood in any province.

Rights to move and gain livelihood

(3) The rights specified in subsection (2) are subject to
(a) any laws or practices of general application in force in a province other than those that discriminate among persons primarily on the basis of province of present or previous residence; and
(b) any laws providing for reasonable residency requirements as a qualification for the receipt of publicly provided social services.

Limitation

(4) Subsections (2) and (3) do not preclude any law, program or activity that has as its object the amelioration in a province of conditions of individuals in that province who are socially or economically disadvantaged if the rate of employment in that province is below the rate of employment in Canada.

Affirmative action programs

Legal Rights

7. Everyone has the right to life, liberty and security of the person and the right not to be deprived thereof except in accordance with the principles of fundamental justice.

Life, liberty and security of person

Search and seizure

8. Everyone has the right to be secure against unreasonable search or seizure.

Detention or imprisonment

9. Everyone has the right not to be arbitrarily detained or imprisoned.

Arrest or detention

10. Everyone has the right on arrest or detention
(*a*) to be informed promptly of the reasons therefor;
(*b*) to retain and instruct counsel without delay and to be informed of that right; and
(*c*) to have the validity of the detention determined by way of *habeas corpus* and to be released if the detention is not lawful.

Proceedings in criminal and penal matters

11. Any person charged with an offence has the right
(*a*) to be informed without unreasonable delay of the specific offence;
(*b*) to be tried within a reasonable time;
(*c*) not to be compelled to be a witness in proceedings against that person in respect of the offence;
(*d*) to be presumed innocent until proven guilty according to law in a fair and public hearing by an independent and impartial tribunal;
(*e*) not to be denied reasonable bail without just cause;
(*f*) except in the case of an offence under military law tried before a military tribunal, to the benefit of trial by jury where the maximum punishment for the offence is imprisonment for five years or a more severe punishment;
(*g*) not to be found guilty on account of any act or omission unless, at the time of the act or omission, it constituted an offence under Canadian or international law or was criminal according to the general principles of law recognized by the community of nations;
(*h*) if finally acquitted of the offence, not to be tried for it again and, if finally found guilty and punished for the offence, not to be tried or punished for it again; and
(*i*) if found guilty of the offence and if the punishment for the offence has been varied between the time of commission and the time of sentencing, to the benefit of the lesser punishment.

12. Everyone has the right not to be subjected to any cruel and unusual treatment or punishment.

Treatment or punishment

13. A witness who testifies in any proceedings has the right not to have any incriminating evidence so given used to incriminate that witness in any other proceedings, except in a prosecution for perjury or for the giving of contradictory evidence.

Self-crimination

14. A party or witness in any proceedings who does not understand or speak the language in which the proceedings are conducted or who is deaf has the right to the assistance of an interpreter.

Interpreter

Equality Rights

15. (1) Every individual is equal before and under the law and has the right to the equal protection and equal benefit of the law without discrimination and, in particular, without discrimination based on race, national or ethnic origin, colour, religion, sex, age or mental or physical disability.

Equality before and under law and equal protection and benefit of law

(2) Subsection (1) does not preclude any law, program or activity that has as its object the amelioration of conditions of disadvantaged individuals or groups including those that are disadvantaged because of race, national or ethnic origin, colour, religion, sex, age or mental or physical disability.
[Note: This section became effective on April 17, 1985. See subsection 32(2) and the note thereto.]

Affirmative action programs

Official Languages of Canada

16. (1) English and French are the official languages of Canada and have equality of status and equal rights and privileges as to their use in all institutions of the Parliament and government of Canada.

Official languages of Canada

(2) English and French are the official languages of New Brunswick and have equality of status and equal rights and privileges as to their use in all institutions of the legislature and government of New Brunswick.

Official languages of New Brunswick

Advancement of status and use

(3)   Nothing in this Charter limits the authority of Parliament or a legislature to advance the equality of status or use of English and French.

Proceedings of Parliament

17. (1) Everyone has the right to use English or French in any debates and other proceedings of Parliament.

Proceedings of New Brunswick legislature

(2)   Everyone has the right to use English or French in any debates and other proceedings of the legislature of New Brunswick.

Parliamentary statutes and records

18. (1) The statutes, records and journals of Parliament shall be printed and published in English and French and both language versions are equally authoritative.

New Brunswick statutes and records

(2)   The statutes, records and journals of the legislature of New Brunswick shall be printed and published in English and French and both language versions are equally authoritative.

Proceedings in courts established by Parliament

19. (1) Either English or French may be used by any person in, or in any pleading in or process issuing from, any court established by Parliament.

Proceedings in New Brunswick courts

(2)   Either English or French may be used by any person in, or in any pleading in or process issuing from, any court of New Brunswick.

Communications by public with federal institutions

20. (1) Any member of the public in Canada has the right to communicate with, and to receive available services from, any head or central office of an institution of the Parliament or government of Canada in English or French, and has the same right with respect to any other office of any such institution where

  (a) there is a significant demand for communications with and services from that office in such language; or

  (b) due to the nature of the office, it is reasonable that communications with and services from that office be available in both English and French.

Communications by public with New Brunswick institutions

(2)   Any member of the public in New Brunswick has the right to communicate with, and to receive available services from, any office of an institution of the legislature or government of New Brunswick in English or French.

21. Nothing in sections 16 to 20 abrogates or derogates from any right, privilege or obligation with respect to the English and French languages, or either of them, that exists or is continued by virtue of any other provision of the Constitution of Canada.

Continuation of existing constitutional provisions

22. Nothing in sections 16 to 20 abrogates or derogates from any legal or customary right or privilege acquired or enjoyed either before or after the coming into force of this Charter with respect to any language that is not English or French.

Rights and privileges preserved

Minority Language Educational Rights

23. (1) Citizens of Canada

(*a*) whose first language learned and still understood is that of the English or French linguistic minority population of the province in which they reside, or

(*b*) who have received their primary school instruction in Canada in English or French and reside in a province where the language in which they received that instruction is the language of the English or French linguistic minority population of the province,

have the right to have their children receive primary and secondary school instruction in that language in that province.

Language of instruction

(2) Citizens of Canada of whom any child has received or is receiving primary or secondary school instruction in English or French in Canada, have the right to have all their children receive primary and secondary school instruction in the same language.

Continuity of language instruction

(3) The right of citizens of Canada under subsections (1) and (2) to have their children receive primary and secondary school instruction in the language of the English or French linguistic minority population of a province

(*a*) applies wherever in the province the number of children of citizens who have such a right is sufficient to warrant the provision to them out of public funds of minority language instruction; and

(*b*) includes, where the number of those children so warrants, the right to have them receive that instruction in

Application where numbers warrant

minority language educational facilities provided out of public funds.

Enforcement

Enforcement of guaranteed rights and freedoms

24. (1) Anyone whose rights or freedoms, as guaranteed by this Charter, have been infringed or denied may apply to a court of competent jurisdiction to obtain such remedy as the court considers appropriate and just in the circumstances.

Exclusion of evidence bringing administration of justice into disrepute

(2) Where, in proceedings under subsection (1), a court concludes that evidence was obtained in a manner that infringed or denied any rights or freedoms guaranteed by this Charter, the evidence shall be excluded if it is established that, having regard to all the circumstances, the admission of it in the proceedings would bring the administration of justice into disrepute.

General

Aboriginal rights and freedoms not affected by Charter

25. The guarantee in this Charter of certain rights and freedoms shall not be construed so as to abrogate or derogate from any aboriginal, treaty or other rights or freedoms that pertain to the aboriginal peoples of Canada including

(*a*) any rights or freedoms that have been recognized by the Royal Proclamation of October 7, 1763; and

(*b*) *any rights or freedoms that may be acquired by the aboriginal peoples of Canada by way of land claims settlement.*

(*b*) any rights or freedoms that now exist by way of land claims agreements or may be so acquired.

[Note: Paragraph 25(b) (in italics) was repealed and the new paragraph substituted by the *Constitution Amendment Proclamation, 1983* (No. 46 *infra*).]

Other rights and freedoms not affected by Charter

26. The guarantee in this Charter of certain rights and freedoms shall not be construed as denying the existence of any other rights or freedoms that exist in Canada.

27. This Charter shall be interpreted in a manner consistent with the preservation and enhancement of the multicultural heritage of Canadians.

28. Notwithstanding anything in this Charter, the rights and freedoms referred to in it are guaranteed equally to male and female persons.

29. Nothing in this Charter abrogates or derogates from any rights or privileges guaranteed by or under the Constitution of Canada in respect of denominational, separate or dissentient schools.

30. A reference in this Charter to a province or to the legislative assembly or legislature of a province shall be deemed to include a reference to the Yukon Territory and the Northwest Territories, or to the appropriate legislative authority thereof, as the case may be.

31. Nothing in this Charter extends the legislative powers of any body or authority.

Application of Charter

32. (1) This Charter applies
(*a*) to the Parliament and government of Canada in respect of all matters within the authority of Parliament including all matters relating to the Yukon Territory and Northwest Territories; and
(*b*) to the legislature and government of each province in respect of all matters within the authority of the legislature of each province.

(2) Notwithstanding subsection (1), section 15 shall not have effect until three years after this section comes into force.
[Note: This section came into force on April 17, 1982. See the proclamation of that date (No. 45 *infra*).]

33. (1) Parliament or the legislature of a province may expressly declare in an Act of Parliament or of the legislature, as the case may be, that the Act or a provision thereof shall operate notwithstanding a provision included in section 2 or sections 7 to 15 of this Charter.

Operation of exception

(2) An Act or a provision of an Act in respect of which a declaration made under this section is in effect shall have such operation as it would have but for the provision of this Charter referred to in the declaration.

Five year limitation

(3) A declaration made under subsection (1) shall cease to have effect five years after it comes into force or on such earlier date as may be specified in the declaration.

Re-enactment

(4) Parliament or the legislature of a province may re-enact a declaration made under subsection (1).

Five year limitation

(5) Subsection (3) applies in respect of a re-enactment made under subsection (4).

Citation

Citation

34. This Part may be cited as the *Canadian Charter of Rights and Freedoms*.

INDEX

LITERARY CREDITS

Chapter 1

Pp. 2–3, 7–8, quotations excerpted from Manning, Morris (1983). *Rights, freedoms and the courts: a practical analysis of the Constitution Act, 1982.* Toronto: Emond-Montgomery. Reprinted by permission of Emond Montgomery.

Pp. 6–7, 9, quotations excerpted from McDonald, David C. (1982) and (1989) editions. *Legal rights in the Canadian Charter of Rights and Freedoms: a manual of issues and sources.* Toronto: Carswell. Reprinted by permission of Carswell, a division of Thomson Canada Limited.

Chapter 2

Pp. 16, quotation excerpted from Manning, Morris (1983). *Rights, freedoms and the courts: a practical analysis of the Constitution Act, 1982.* Toronto: Emond-Montgomery. Reprinted by permission of Emond Montgomery.

Chapter 3

Pp. 61, 68–69, 82, quotations excerpted from Salhany, Roger E. (1997). *Police manual of arrest, seizure and interrogation.* 7th ed. Scarborough, Ont.: Carswell. Reprinted by permission of Carswell, a division of Thomson Canada Limited.

P. 63, quotation from Manning, Morris (1993). *Rights, freedoms and the courts:* a practical analysis of the *Constitution Act,* 1982. Toronto: Emond-Montgomery. Reprinted by permission of Emond-Montgomery.

Chapter 4

Pp. 92–93, quotation excerpted from Delisle, Ronald Joseph and Don Stuart (1996). *Learning Canadian Criminal Procedure.* 4th ed. Toronto: Thomson Canada. Reprinted by permission of Carswell, a division of Thomson Canada Limited.

Pp. 93, 94, 95, 97, 98, 99, 100–101, 106–107, quotations excerpted from Trotter, Gary T. (1992). *The Law of Bail in Canada.* Scarborough, Ont.: Carswell. Reprinted by permission of Carswell, a division of Thomson Canada Limited.

Chapter 5

Pp. 124, 126, quotations excerpted from Sheehan, Robert and Gary W. Cordner (1989). *Introduction to Police Administration.* 2nd edition. Cincinatti, Ohio: Anderson Publishing Co. Reprinted by permission of Anderson Publishing Co.

Pp. 121, 123, 126–128, 129, 131, quotations reprinted by permission of the publisher from *Varieties of Police Behavior: The Management of Law and Order in Eight Communities* by James Q. Wilson, Cambridge, Mass.: Harvard University Press, Copyright © 1968, 1978 by the President and Fellows of Harvard College.